ELEMENTARY NUMBER THEORY

ELEMENTARY NUMBER THEORY

EFRAIM P. ARMENDARIZ

STEPHEN McADAM

University of Texas at Austin

MACMILLAN PUBLISHING CO., INC.

New York

Collier Macmillan Publishers

London

MACMILLAN PUBLISHING CO., INC.
866 Third Avenue, New York, New York 10022

COLLIER MACMILLAN CANADA, LTD.

Library of Congress Cataloging in Publication Data

Armendáriz, Efraim P
Elementary number theory.

Includes index.
1. Numbers, Theory of. I. McAdam, Stephen, joint author. II. Title.
QA241.A73 1980 512′.72 79-15731
ISBN 0-02-303850-0

Printing: 1 2 3 4 5 6 7 8 Year: 0 1 2 3 4 5 6

To the Memory of Wilson

PREFACE

To the Student

In writing this book, we have tried to make the reading of the mathematics involved as easy and pleasant as possible. Should you elect to put in the necessary effort, you will be amply rewarded. Because reading mathematics is quite different from reading cereal boxes, newspapers, or history books, we now offer a few suggestions concerning how to go about it.

1. Be patient. Do not expect to be able to read and easily understand more than a few pages at a time. Often, reading mathematics will require such a pace, and to read it any faster accomplishes almost nothing, for you will understand very little.

2. Read and reread. To understand much of mathematics often requires reading things two or three times. Thus it helps to read a sentence or statement twice to ensure that you understand its details. Read sections twice to be certain that you grasp the main details. Look over earlier sections to refresh your memory.

3. Know what you are talking about. You cannot read mathematics unless you know exactly what the words and symbols mean. When you encounter a new word, phrase, or symbol, stop and learn precisely what it means.

4. Ask questions. Certainly, you should ask questions of your instructor, if it will help you. But more important, ask them of yourself. What does that word mean? Do I remember exactly what that theorem said? How does this theorem use the previous results? How might this theorem be used later?

5. Think about what you read. This is really the same as 4, but it is worth rephrasing. Remember, you can be our own best teacher.

6. Be patient and persistent, keep the definitions in mind, ask yourself what you are attempting and remember what knowledge you have to work with. You can succeed.

To the Instructor—*The Spirit of the Book*

We hope you find this text useful. Our intent in writing it was to provide a vehicle by which a mathematically unsophisticated student with no preparation other than basic high school algebra could be introduced to rigorous, nontrivial mathematics. Our choice of elementary number theory for this introduction is based on several fairly obvious facts. Elementary number theory is concrete, dealing with the integers rather than an abstract mathematical system; thus the students can hone their abilities at rigorous thinking in a familiar setting. The subject matter is accessible to anyone having a minimum of high school algebra. It uses techniques common to all of mathematics. Finally, it contains results as lovely and refined as any in mathematics.

On the basis of the preceding paragraph we may be accused of using elementary number theory as a soapbox from which to preach mathematics. Other evidence supporting this charge will be found in the text. At times we have included presentations of two proofs to a given result, suggesting that the cognative processes involved are of even more importance than the theorems. The evidence is corroborated by our discussions and questions concerning the relative merits of proofs, definitions, and techniques. To this and any similar accusations, we can only smile and hope we are found guilty. You are the judge and the students are the jury.

We have already mentioned that we assume no mathematical maturity of readers starting this book. Specifically, in the early chapters we present algebraic manipulations in complete detail, we discuss the purpose of definitions and of symbolism, and we point out that proofs often use standard arguments such as considering various cases or arguing by contradiction. The text is rife with examples, and we repeatedly stress the importance of looking at examples when formulating a plausible conjecture or fashioning a proof. We also include numerous historical remarks, not only for their intrinsic interest, but also to present an image of mathematics as a continuing effort of the human spirit, an image fostered by our mention of problems still under investigation. Finally, throughout we attempt to instill in the reader a sense of confidence, a feeling that this is material which can be mastered and enjoyed.

A few words are in order concerning our choice of certain options. Since a familiarity with the principles of induction and/or well-ordering is soon essential to a rigorous treatment of the material, we choose to present these in Chapter I, rather than banishing them to an appendix. We present both induction and well-ordering rather than just one of these, again from a sense of realism. Both are commonly used in mathematics. We feel that the standard statement of well-ordering, "a nonempty set of integers bounded below, contains a least element," is beyond the grasp of most beginners. Our statement of it is, we hope, much more obvious. In Chapter 2 we introduce ideals as the tool used to solve $aX + bY = c$ and to discuss GCDs, relegating the Euclidean algorithm to a brief discussion at the end. We favor ideals not only because they appear in other areas of mathematics, but also because they provide an excellent example of how an abstract notion can have very concrete applications. In Chapter 6 we entirely answer the question of when a system of linear congruences has a solution. We do this primarily to satisfy our own sense of completeness. However, matters are arranged so that those wishing to only present the Chinese Remainder Theorem can, indeed, do so quickly, ignoring the rest.

We have also exercised numerous options regarding topics left out of this text. Our underlying desire was to present a few of the most important (and beautiful) topics with love and patience, even at the cost of having to omit some worthy ideas. In particular, we have not included such topics as writing numbers in various bases, tests for divisibility, or casting out nines. While these are legitimate topics that are often included in number theory texts aimed at the audience we hope to reach, we feel that they lack the sophistication found in the ideas we do present. Although we proceed slowly and carefully, we nonetheless strive to grasp the highest level of mathematics within our reach.

Each chapter ends with a plethora of exercises, arranged in two groups. Those in group A are certainly accessible to all students. To a large extent, they reinforce understanding of the material by asking for numerical verification of the various results. They also require proofs at an elementary level. The B exercises are more diversified. Some explore comments made in the text, some indicate directions the theory might have taken, and some are merely fun. Some of the B exercises can be done by any student, while others will challenge the best student.

The following table describes the prerequisites of each chapter in the book.

Chapter	Prerequisite
1	High school algebra
2	1
3	1,2
4	1,2,3
5	1–4
6	1–4
7	1–4
8	1–4, 5 (5.1), 6 (6.3, 6.4)
9	1–4, 6 (6.4)

We acknowledge the many students who have patiently and dutifully used this text in its preliminary versions. Their comments and concerns were taken into consideration in the final preparation. Our indebtedness to two excellent typists is great. Nita Goldrick prepared preliminary versions with artistic skill and efficiency, while the final touch-ups were gracefully applied by Suzy Crumley. We thank John Goth for compiling the tables of primes and squares. A special note of gratitude is extended to Edna Armendáriz and Martha Smith for their continual support and encouragement for not only this venture but for many others as well.

CONTENTS

CHAPTER **1**	**Rules of the Game**	1
	1.1 The Integers	1
	1.2 Well-Ordering	2
	1.3 Induction	6
	EXERCISES	7
CHAPTER **2**	**Divisibility**	10
	2.1 Division	10
	2.2 Greatest Common Divisors	22
	2.3 The Equation $aX + bY = c$	21
	EXERCISES	30
CHAPTER **3**	**Prime Numbers**	33
	3.1 Primes	33
	3.2 A Sieve Method	39
	3.3 Fermat Factorization	40
	3.4 Some Unsolved Problems	41
	EXERCISES	42
CHAPTER **4**	**Congruences**	45
	4.1 Congruence	45
	4.2 Inverses Modulo m	49
	4.3 Fermat's Little Theorem	53
	4.4 Euler's Theorem	55
	4.5 A Historical View	58
	4.6 Wilson's Theorem	60
	EXERCISES	62
CHAPTER **5**	**Multiplicative Functions**	66
	5.1 Three Special Functions	66
	5.2 Perfect Numbers	71
	5.3 Möbius Inversion	74
	EXERCISES	79

CHAPTER **6** **Solving Congruences** 82

6.1 The Congruence $aX \equiv b \bmod m$ 83

6.2 Systems of Linear Congruences 85

6.3 The Congruence $X^2 \equiv a \bmod m$ 101

6.4 A General Theorem 111

EXERCISES 113

CHAPTER **7** **Sums of Squares** 119

7.1 Pythagorean Triples 123

7.2 Sums of Two Squares 128

7.3 Sums of Four Squares 134

EXERCISES 139

CHAPTER **8** **Primitive Roots** 143

EXERCISES 151

CHAPTER **9** **Quadratic Reciprocity** 153

9.1 The Legendre Symbol 153

9.2 The Quadratic Reciprocity Theorem 158

EXERCISES 175

TABLE OF SQUARES 177

TABLE OF PRIMES LESS THAN 10,000 178

INDEX 191

1

Rules of the Game

NUMBER theory is the study of the multiplicative and additive properties of the integers. You already know how to add and multiply integers, but our study will reveal that integers behave in a manner far more intriguing than you may have imagined. An example of this is the fact that every positive integer can be expressed as the sum of four squares. For instance,

$$62 = 6^2 + 5^2 + 1^2 + 0^2$$
$$63 = 6^2 + 5^2 + 1^2 + 1^2$$
$$64 = 8^2 + 0^2 + 0^2 + 0^2$$

You can verify this property for other integers; the table of squares at the end of the book would be helpful. You might also wonder why this statement is true for *all* positive integers. This will be one of the many things we will demonstrate, and is one of the topics considered in Chapter 7. Before we can jump into trying to establish this or other sophisticated properties of integers, we must agree on certain basic assumptions. This chapter is designed to help us do that. First we summarize properties of addition, multiplication, and order that we will assume as fact. Then we will examine two properties of the integers that distinguish this system from other number systems. Much of what is presented here should be familiar to you, and therefore we will be brief. Let us point out, however, that it took the human race thousands of years to realize that the rules given here govern the behavior of the integers. Moreover, the consequences of these simple rules are far richer than is immediately obvious, as we will demonstrate in subsequent chapters. Best of all, not all consequences are fully known or understood, so that much exploration and discovery remain for those with an inquisitive nature.

1.1 The Integers

The letter $\mathbb{Z}$ is used to denote the set of integers

$$\mathbb{Z} = \{\ldots, -5, -4, -3, -2, -1, 0, 1, 2, 3, 4, 5, \ldots\}$$

You should be familiar with most of the properties of this set involving addition and multiplication. For instance, you know that the members of this set obey all of the following rules:

1° $a + b = b + a$ and $ab = ba$.
2° $(a + b) + c = a + (b + c)$ and $a(bc) = (ab)c$.
3° $0 + a = a$ and $1a = a$.
4° $-a = (-1)a$ and $a - a = a + (-a) = 0$.
5° $a(b + c) = ab + ac$.
6° $0a = 0$ and if $ab = 0$, then either $a = 0$ or $b = 0$.

where a, b, and c represent any members of the set $\mathbb{Z}$.

There is also an ordering on the members of $\mathbb{Z}$, represented by the symbol "$<$" and read "less than," which, as you probably know, has the following properties:

7° For any $a \in \mathbb{Z}$, if $a \neq 0$, then either $a < 0$ or $0 < a$.
8° If $a < b$ and $b < c$, then $a < c$.
9° If $a < b$, then $a + c < b + c$ for any integer c.
10° If $a < b$ and $c < 0$, then $bc < ac$.
11° If $a < b$ and $0 < c$, then $ac < bc$.

From these properties you can derive other properties of the integers,

EXAMPLES

(a) It is true that for all integers x, y, and z,

$$-(x + y) = (-x) + (-y)$$

Indeed,

$$\begin{aligned} -(x + y) &= (-1)(x + y) && \text{(Property 4°)} \\ &= (-1)x + (-1)y && \text{(Property 5°)} \\ &= (-x) + (-y) && \text{(Property 4°)} \end{aligned}$$

(b) It is also true that for any integer $x \neq 0$, $x^2 > 0$. Here is why.

STEP 1: $x \neq 0$, so $x < 0$ or $0 < x$ (Property 7°)

STEP 2: If $x < 0$, then $0 \cdot x < x \cdot x$ (Property 10° with $a = x$, $b = 0$, and $c = x$)

$0 < x^2$ (Property 6°)

STEP 3: If $0 < x$, then $0 \cdot x < x \cdot x$ (Property 11°)

$0 < x^2$ (Property 6°)

1.2 Well-Ordering

So far, we have told you nothing about the set $\mathbb{Z}$ that you did not already know. Nor have we told you anything about the set $\mathbb{Z}$ that does not hold equally well for the set of rational numbers or for the set of real numbers, since the statements 1° through 11° remain true if we allow a, b, and c to represent either rational numbers or real numbers. Our desire in this chapter, then, is to present and discuss two properties of $\mathbb{Z}$ which neither the set of rational numbers nor the set of real numbers can claim. The first distinguishing property that we are speaking about and will present shortly is called the *Principle of Well-Ordering* and in a way it will be of greater importance to us than any of the properties 1° through 11°. For instance, suppose that we are able to prove a theorem about integers using only the properties 1° through 11°; such a theorem might be: "For all integers a, b, and c we have $a(b + c) = ba + ca$." (The truth of this statement is easily established by combining 5° with 1°.) Since the members of the set of rational numbers also obey rules 1° through 11°, the proof works equally well for rational numbers as for integers, so $a(b + c) = ba + ca$ is true regardless of whether we take a, b, and c to be integers or rational numbers (or real numbers for that matter). On the other hand, the Principle of Well-Ordering, which we have not yet stated, holds only for $\mathbb{Z}$ and not for the rational numbers or the real numbers. Therefore, if we prove a theorem about integers using, as one step in our proof, the Principle of Well-Ordering, that theorem is only valid for $\mathbb{Z}$ and not for the real numbers or the rational numbers (unless by lucky coincidence!).

Since we have given this principle such a big buildup, and since it has such a fancy name, you are probably expecting something very exotic. Not so; it is quite simple. Consider the following three subsets of $\mathbb{Z}$:

$$A = \varnothing \text{ (the empty set, the set having no members)}$$
$$B = \{a \in \mathbb{Z} \mid a > 7\} = \{8, 9, 10, 11, \ldots\}$$
$$C = \{a \in \mathbb{Z} \mid a \text{ is even}\} = \{\ldots, -6, -4, -2, 0, 2, 4, 6, \ldots\}$$

Observe that A contains no integers, B contains a smallest integer (namely 8), and C contains arbitrarily small integers. In the case of C, we mean that if you choose any number there is always a number in C that is less than the one you have chosen: Does C contain a number less than -20? Yes, -22. Does C contain a number less than $-1{,}000{,}001$? Yes, $-1{,}000{,}002$. The Principle of Well-Ordering asserts that any subset of $\mathbb{Z}$ behaves exactly like one of A, B, or C; that is:

The Principle of Well-Ordering. If D is a subset of $\mathbb{Z}$, then either

(a) D contains no integers;

or (b) D contains a smallest integer;

or (c) D contains arbitrarily small integers.

Notice that the three statements (a), (b), and (c) are mutually exclusive so that, for example, no set of integers can simultaneously satisfy both (a) and (c). You are encouraged to classify the following sets according to type (a), type (b), or type (c):

1. $\{a \in \mathbb{Z} \mid a^2 \text{ is positive}\}$.
2. $\{a \in \mathbb{Z} \mid a^2 \text{ is negative}\}$.
3. $\{a \in \mathbb{Z} \mid 2a + 1 > 6\}$.
4. $\{a \in \mathbb{Z} \mid 3a = 4\}$.

If you feel that the Principle of Well-Ordering is a fairly obvious observation concerning the integers, you are right! But keep in mind that the Principle applies only to $\mathbb{Z}$, because the set of rational numbers (or the set of real numbers) does not satisfy the Principle. For example, consider the following set of rational numbers:

$$S = \{2 + 10^{-m} \mid m = 1, 2, 3, \ldots\} = \{2.1, 2.01, 2.001, \ldots\}$$

Certainly, S contains rational numbers, so it is not of type (a). Does S contain a smallest rational number? The answer is no because $2 + 10^{-(m+1)} < 2 + 10^{-m}$ for any positive integer m. Thus S is not of type (b). Finally, does S contain arbitrarily small rational numbers? Again the answer is no, because every number in S is greater than 2. Thus S is not of type (c). The reason for this fact, of course, is that S is not a set consisting only of integers.

We would like to demonstrate one way in which the Principle of Well-Ordering can be used. As an example, let us prove the statement:

$$(*) \qquad \text{If } n \in \mathbb{Z},\ n \geq 1, \quad \text{then} \quad 1 + 2 + \cdots + n = \frac{n(n+1)}{2}$$

For the proof, let $D = \{n \in \mathbb{Z} \mid n \geq 1 \text{ and } 1 + 2 + \cdots + n \neq n(n+1)/2\}$. That is, D consists of all those integers greater than or equal to 1 for which the statement $(*)$ is false. As we hope to show that the statement is never false, our goal is to show that D is empty.

By the Principle of Well-Ordering we know that either D is empty (as we hope), or D contains a least integer, or D contains arbitrarily small integers. If we can eliminate these last two possibilities, our hopes are realized; only the first possibility remains, D is empty, and statement $(*)$ is true for all positive integers.

The possibility that D contains arbitrarily small integers is easily eliminated. After all, by the definition of the set D, if $n \in D$, then $n \geq 1$. Let us now proceed to eliminate the possibility that D contains a smallest integer. We will do this by "contradiction." That is, we will assume that D does contain a smallest integer and show that this assumption leads to a logical contradiction and must therefore be false.

Assume that D contains a smallest integer and call it d. Then since $d \in D$, we have $d \geq 1$ and $1 + 2 + \cdots + d \neq d(d+1)/2$. Clearly, $d \neq 1$ since

$1 = 1(1 + 1)/2$. Therefore, $d > 1$. Consider $d - 1$. Since $d > 1$, $d - 1 \geq 1$. Since $d - 1 < d$ and d is the smallest integer in the set D, $d - 1 \notin D$. Therefore, the statement (*) is true when $n = d - 1$, so $1 + 2 + \cdots + (d - 1) = (d - 1)(d - 1 + 1)/2$. Now add d to both sides of this equality. On the left-hand side, we get $1 + 2 + \cdots + (d - 1) + d$ and on the right-hand side we get

$$\begin{aligned}\frac{(d-1)(d-1+1)}{2} + d &= \frac{(d-1)d}{2} + d = \frac{(d-1)d}{2} + \frac{2d}{2} \\ &= \frac{(d-1)d + 2d}{2} = \frac{d^2 - d + 2d}{2} \\ &= \frac{d^2 + d}{2} = \frac{d(d+1)}{2}\end{aligned}$$

Therefore, our original equality $1 + 2 + \cdots + (d - 1) = (d - 1)(d - 1 + 1)/2$ leads to the equality $1 + 2 + \cdots + d = d(d + 1)/2$. However, this cannot be true because $d \in D$ and hence $1 + 2 + \cdots + d \neq d(d + 1)/2$. This is our desired contradiction; it establishes that the assumption that D has a least element is indeed false. Therefore, we have eliminated all possibilities except that D is empty, as desired.

We have given the proof in great, and perhaps exhausting, detail. In order to keep this book from quadrupling in length as well as in price· we will be much more brief in the future, leaving it to you to recognize standard techniques such as considering various cases, using contradiction, or doing a certain amount of simple algebraic manipulations. In the interest of pedagogy and to clarify the previous statement, we repeat the foregoing proof in the more traditional mathematical style.

Theorem 1.1. For $n \in \mathbb{Z}$ and $n \geq 1$, $1 + 2 + \cdots + n = n(n + 1)/2$.

Proof. Let $D = \{n \in \mathbb{Z} \mid n \geq 1 \text{ and } 1 + 2 + \cdots + n \neq n(n + 1)/2\}$. We must show $D = \varnothing$. Since for any $n \in D$, we have $n \geq 1$, by the Principle of Well-Ordering, we have only to show that D does not contain a least integer. Suppose that it does and call it d. Then $d > 1$ since $1 = 1(1 + 1)/2$. Now consider $d - 1$. Then $1 \leq d - 1 < d$ and since d is the least integer in D, $d - 1 \notin D$. Thus $1 + 2 + \cdots + (d - 1) = (d - 1)(d - 1 + 1)/2$. Adding d to both sides of this equality results in the equality

$$\begin{aligned}1 + 2 + \cdots + (d - 1) + d &= (d - 1)(d - 1 + 1)/2 + d \\ &= [(d - 1)d + 2d]/2 = d(d + 1)/2\end{aligned}$$

which contradicts the fact that $d \in D$. ∎

Comparing the two presentations of this proof, you can see that the second one leaves out laborious details of standard techniques.

1.3 Induction

Next, we wish to acquaint you with another very useful standard technique (if you are not already aware of it). It is called the *Principle of Induction.* After presenting it, we will use it to give another proof of Theorem 1.1; the proof is more appealing than that just given.

The Principle of Induction. Suppose that S is a subset of the integers and that d is a fixed integer. Suppose further that the following two conditions hold:

(i) $d \in S$.

(ii) For any integer $x \geq d$, if $\{d, d+1, d+2, \ldots, x\} \subseteq S$, then $x + 1 \in S$.

Then every integer greater than or equal to d belongs to S.

It might be a good idea to examine examples of some sets S so as to get a good grasp of what the principle states, especially (ii). For example, let $S = \{-7, -6, -5, -3, -2, -1, \ldots\}$ together with $d = -7$. The set S fails to satisfy condition (ii) since if we let $x = -5$, then $\{-7, -6, -5\} \subseteq S$ but $x + 1 = -4 \notin S$. As a second example, let $S = \{-3, -2, -1, 0, 1, 2, \ldots\}$ (assuming that by the dots we mean the rest of the integers after 2) and let $d = -3$. Then this set S does obey condition (ii). For example, $\{-3, -2\} \subseteq S$ so that (ii) demands that $-2 + 1 = -1$ belong to S, which is certainly true. Similarly, $\{-3, -2, -1, 0, 1, 2, 3\} \subseteq S$, so (ii) insists that $4 = 3 + 1 \in S$, as is the case.

Here is a general view of induction. Suppose that we have a set S and an integer d which satisfy conditions (i) and (ii). Is $d \in S$? Yes, because (i) tells us that $d \in S$. Is $d + 1 \in S$? Well, we already know that $d \in S$, so let us apply condition (ii) to the subset $\{d\}$. Since $\{d\} \subseteq S$, condition (ii) tells us that $d + 1 \in S$. Therefore, $d, d + 1$ belong to S. Is $d + 2 \in S$? Since we know that $\{d, d + 1\} \subseteq S$, condition (ii) again ensures that $d + 2 = (d + 1) + 1 \in S$. Again we ask, is $d + 3 \in S$? We already know that $\{d, d + 1, d + 2\} \subseteq S$, so (ii) insists that $d + 3 = (d + 2) + 1 \in S$. Continuing in this manner we see that $\{d, d + 1, d + 2, d + 3, \ldots\} \subseteq S$; that is, every integer greater than or equal to d belongs to S.

As we promised, we can use induction to handle things such as Theorem 1.1.

Theorem 1.1 (Revisited). If $n \in \mathbb{Z}, n \geq 1$, then $1 + 2 + \cdots + n = n(n + 1)/2$.

Proof. Let $S = \{n \in \mathbb{Z} \mid n \geq 1 \text{ and } 1 + 2 + \cdots + n = n(n + 1)/2\}$ and let $d = 1$. Since $1 = 1(1 + 1)/2$ we have $d = 1 \in S$ and hypothesis (i) of the Principle of Induction is satisfied. To show that (ii) holds, assume that $x \geq 1$ and that $\{1, 2, \ldots, x\} \subseteq S$. (In this instance we will not need the full force of the inclusion but only that $x \in S$; later applications of induction will require using the fact that all integers between d and x lie in S.) Since $x \in S$, we have $1 + 2 + \cdots + x = x(x + 1)/2$. Adding $x + 1$ to both sides leads to

$$\begin{aligned} 1 + 2 + \cdots + x + (x + 1) &= x(x + 1)/2 + (x + 1) = [x(x + 1) + 2(x + 1)]/2 \\ &= [(x + 1)(x + 2)]/2 = (x + 1)[(x + 1) + 1]/2 \end{aligned}$$

which shows that $x + 1 \in S$. Hence $\{1, 2, \ldots, x, x + 1\} \subseteq S$. Accordingly, by the Principle of Induction, for all integers $n \geq 1$, $n \in S$. ∎

Well-ordering and induction are first cousins. More precisely, it is a fact that any result which can be proved using well-ordering can also be proved using induction, and vice versa. (Mathematicians say that the two principles are equivalent.) Why, then, do we need both? The answer is: for convenience. For sometimes a result can be proved more slickly using one of these principles rather than the other. For instance, we think you will agree that the inductive proof we gave of Theorem 1.1 is nicer than the proof using well-ordering. Only experience can tell you when to use one of these principles and which is more efficient at the time.

In the coming chapters you will see many uses of both induction and well-ordering. Let us mention that this book is not about these principles. But rather these principles are vital tools needed to study what the book is about, number theory.

EXERCISES 1.A

1. Using Properties 1° through 6°, verify that the following statements are true (x, y, and z represent integers).
(a) $x + (z + y) = (x + y) + z$
(b) $xy + yz = y(x + z)$
(c) $x(y - z) = xy - xz$
(d) If $x + y = x + z$, then $y = z$.
(e) If $x \geq 1$, $y \geq 1$ and $xy = z$, then $z \geq x$ and $z \geq y$. (*Hint*: Assume that $x > z$ and deduce that $z = xy > zy > z$, a contradiction.)

2. Using Properties 7° through 11°, verify that the following statements are true (x, y, and z represent integers.)
(a) If $x < y$, then $0 < y - x$.
(b) If $0 < y - x$, then $x < y$.
(c) If $x + z < y + z$, then $x < y$.
(d) $0 < 1$

3. Find the smallest integer in each of the following sets.
(a) $A = \{2k \mid k \in \mathbb{Z} \text{ and } k \geq -7\}$
(b) $B = \{k \in \mathbb{Z} \mid 13k + 4 > -6\}$
(c) $C = \{k \in \mathbb{Z} \mid k > 0 \text{ and } k = 2n + 3m \text{ for some } n, m \in \mathbb{Z}\}$
(d) $D = \{k \in \mathbb{Z} \mid k > 0 \text{ and } k = 15n + 20m \text{ for some } n, m \in \mathbb{Z}\}$
(e) $E = \{k \in \mathbb{Z} \mid k > 0, k = 6n \text{ and } k = 9m \text{ for some } n, m \in \mathbb{Z}\}$
(f) $F = \{k \in \mathbb{Z} \mid 3^k - 1 = 7n \text{ for some } n \in \mathbb{Z}\}$
(g) $G = \{k \in \mathbb{Z} \mid k > 17 \text{ and } 7k - 2 = 4n \text{ for some } n \in \mathbb{Z}\}$
(h) $H = \{k \in \mathbb{Z} \mid k > 23 \text{ and } k^2 + 1 = 5n \text{ for some } n \in \mathbb{Z}\}$

4. Use the Principle of Induction to show that each of the following statements is true for all positive integers n.
(a) $2 + 4 + 6 + \cdots + 2n = n(n + 1)$
(b) $1 + 4 + 7 + \cdots + (3n - 2) = \dfrac{n(3n - 1)}{2}$
(c) $1^2 + 2^2 + 3^2 + \cdots + n^2 = \dfrac{n(n + 1)(2n + 1)}{6}$
(d) $\dfrac{1}{2} + \dfrac{1}{2^2} + \dfrac{1}{2^3} + \cdots + \dfrac{1}{2^n} = 1 - \dfrac{1}{2^n}$
(e) $\dfrac{1}{1 \cdot 2} + \dfrac{1}{2 \cdot 3} + \dfrac{1}{3 \cdot 4} + \cdots + \dfrac{1}{n(n + 1)} = \dfrac{n}{n + 1}$
(f) $2^n \geq 2n$
(g) $x^n - 1 = (x - 1)(x^{n-1} + x^{n-2} + \cdots + x^2 + x + 1)$ for all real numbers x
(h) $1 + 3 + 5 + \cdots + (2k - 1) = k^2$
(i) $2^n < n!$ if $n \geq 4$ [*Note*: $n! = n(n - 1)(n - 2) \cdots 3 \cdot 2 \cdot 1$).]

5. Show that each of the statements in Exercise 1.A.4 is true by using the Principle of Well-Ordering.

6. Why is the letter $\mathbb{Z}$ used to denote the integers, instead of the letter I? (A knowledge of German is helpful here.)

EXERCISES 1.B

1. Use the Well-Ordering Principle to prove the following statement: If a and b are positive integers, then there exists an integer n such that $b \leq na$.

2. Show that if $a \in \mathbb{Z}$ and $0 < a$, then $1 \leq a$.
3. Show that if D is a nonempty set of integers and there exists an integer x such that $d \leq x$ for all $d \in D$, then D has a largest element; that is, there is an integer m *in* D such that $x \leq m$ for all $x \in D$.
4. In Chapter 2 the following definition is given: *a divides b* if there is an integer c such that $b = ac$. Show that if b is positive, then the set $D = \{a \mid a \neq b \text{ and } a \text{ divides } b\}$ has a largest element.
5. Consider the set S of all ordered pairs of integers (a, b). Give this set an ordering "$\ll$" in the following manner: For any two pairs (a_1, b_1) and (a_2, b_2) let $(a_1, b_1) \ll (a_2, b_2)$ if either $a_1 < a_2$ or if $a_1 = a_2$ and $b_1 < b_2$ [example: $(-3, 3) \ll (4, -7)$ and $(2, 4) \ll (2, 6)$].
   (a) Show that for any (a_1, b_1) and (a_2, b_2), exactly one of the next statements is true:
   $$(a_1, b_1) \ll (a_2, b_2), \qquad (a_2, b_2) \ll (a_1, b_1), \qquad (a_1, b_1) = (a_2, b_2)$$
   (b) Show that if $(a_1, b_1) \ll (a_2, b_2)$ and $(a_2, b_2) \ll (a_3, b_3)$, then $(a_1, b_1) \ll (a_3, b_3)$.
   (c) Show that the Principle of Well-Ordering fails for this set S with this order relation.
6. Prove that the Principle of Induction is a consequence of the Principle of Well-Ordering.
7. Prove that the Principle of Well-Ordering is a consequence of the Principle of Induction.

2
Divisibility

In Chapter 1 we discussed two principles that may have been unfamiliar to you, the Principle of Well-Ordering and the Principle of Induction. Both of these principles might be considered background material and do not truly represent the spirit of the theory of numbers. Rather, they are tools used in its study. The present chapter deals with a topic that is probably quite familiar to you, division. Our goals are to state in a precise manner what division is and what its properties are, as well as to see how much information about integers the concept of division can provide.

2.1. Division

Perhaps we should begin by pointing out that of the four basic operations of addition, subtraction, multiplication, and division, the last of these is probably the most interesting one. We say this because if you take any two integers x and y, then $x + y$, $x - y$, and xy are always integers; however the fraction $\frac{x}{y}$ may or may not be an integer. For instance, $\frac{4}{2}$ is an integer whereas $\frac{5}{2}$ is not.

At first sight, the question of whether a particular fraction $\frac{x}{y}$ is or is not an integer may not seem very exciting. However, examining the two examples above as well as others may convince you that the question has some merit because in order for $\frac{x}{y}$ to be an integer, there must be some intimate connection between x and y. For your further consideration let us make the following choices of x and y: let a be any integer, let $x = a^3 - a$ and let $y = 3$. Now certainly $(a^3 - a) + 3$ is an integer and we would be technically justified in stating a theorem as follows.

"**Theorem.** $(a^3 - a) + 3$ is always an integer for any integer a."

However, number theory, as well as mathematics in general, is more than just a compilation of true facts. The theorem above is true but incredibly dull. Any mathematics text that included the theorem above might not be criticized for its accuracy, but it would surely be criticized for its taste.* On the other hand, a true and extremely interesting fact is:

"**Theorem.** $\frac{a^3 - a}{3}$ is *always* an integer for any integer a."

Since the quotient of two integers is not always an integer, the previous observation might be worth recording and your interest in it might increase once you try to verify it. The verification is not immediate but after completing this chapter you should be able to verify it; this is Exercise 2.A.8. For the present let us give you a few values of a and convince you that the statement just might be true.

a	$a^3 - a$	$\frac{a^3 - a}{3}$
0	0	0
1	0	0
2	6	2
3	24	8
−5	−120	−40
123	1,860,744	620,248

Let us now turn to the task of making precise some of what we know about the process of division. Suppose that we begin by asking, does 3 divide 4? In third grade, where students only work with whole numbers, the answer is no; as the tykes so charmingly put it, "3 into 4 won't go." However, in fourth grade, where fractions are studied, the answer is yes, 3 does divide 4, giving an answer of $\frac{4}{3}$. Thus the answer to the question "Does 3 divide 4?" depends on the frame of reference. If you are working within $\mathbb{Z}$, the integers, then 3 does not divide 4. However, if you are working within Q, the rational numbers, then 3 does divide 4. We see, therefore, that our original question was vague, and this vagueness is due to the fact that our use of the term "divides" was vague. Fortunately, it is simple to remedy this vagueness. Since this book deals exclusively with $\mathbb{Z}$, we will always use "divides" with $\mathbb{Z}$ as our frame of

* We of course exclude the present text.

reference, just like the third-graders. With this in mind we give the following definition.

Definition. If a and b are in $\mathbb{Z}$ and $a \neq 0$, then *a divides b* means that there is an integer c such that $b = ac$.

According to this definition, 3 does not divide 4, since there is no integer which when multiplied by 3 gives 4. It is true of course that $c = \frac{4}{3}$ makes the statement $4 = 3c$ true, but $\frac{4}{3}$ is not an integer. On the other hand, 5 does divide -35 since $-35 = 5(-7)$ and $-7 \in \mathbb{Z}$.

Before proceeding further we mention something about definitions. One way of thinking of them is as a sort of shorthand. The phrase "there is an integer c such that $b = ac$" is relatively long, and if you had to speak it or write it many times the process would become tiresome. Therefore, we agree to use the phrase "a divides b" as a shorthand substitute for the longer phrase, thus saving time, ink, and some mental effort. Thus a definition is a labor-saving device.

While the phrase "a divides b" is acceptably short in speech, in writing even it can become tedious. As mathematicians long have been considered as being among the laziest people in the world, we introduce a shorthand for the shorthand. Instead of writing "a divides b" we write "$a \mid b$"; in speech the symbols "$a \mid b$" are still said "a divides b."

Notation. $a \mid b$ means "a divides b." Also, $a \nmid b$ means "a does not divide b."

Note that this is analogous to use of $a = b$ to mean "a equals b" and $a \neq b$ to mean "a does not equal b."

We return to looking at some of the properties of division. Many of the results in this chapter will be of little interest in their own right, although there will be some exceptions. However, all these results will have numerous applications in what comes later, giving them an importance far beyond their intrinsic value. Some of the results we obtain are called *lemmas*; a lemma usually identifies a result that is used as a tool in later applications rather than a result that is an end in itself.

Lemma 2.1. If $u, v_1, v_2, \ldots, v_n$ are in $\mathbb{Z}$ and $u \mid v_1, u \mid v_2, \ldots, u \mid v_n$, then $u \mid (v_1 + v_2 + \cdots + v_n)$.

Proof. For each $i = 1, 2, \ldots, n$ we are given that $u \mid v_i$. Let us write out explicitly what that shorthand statement means: we are told that for each $i = 1, 2, \ldots, n$ there is an integer c_i such that $v_i = uc_i$. Thus $v_1 = uc_1$, $v_2 = uc_2, \ldots, v_n = uc_n$. Our goal is to show that $u \mid (v_1 + v_2 + \cdots + v_n)$. That is, we wish to find an integer c such that $(v_1 + v_2 + \cdots + v_n) = uc$. However,

$$(v_1 + v_2 + \cdots + v_n) = (uc_1) + (uc_2) + \cdots + (uc_n) = u(c_1 + \cdots + c_n)$$

the last equality resulting from an application of the distributive law (factoring out the common term u). Since $c_1 + c_2 + \cdots + c_n$ is an integer, we can let $c = c_1 + c_2 + \cdots + c_n$. Then $(v_1 + v_2 + \cdots + v_n) = uc$. From the definition we conclude that $u \mid (v_1 + v_2 + \cdots + v_n)$. ∎

Incidentally, the symbol "∎" is used throughout this book to identify the end of a proof.

As simple as the preceding lemma and proof are, they reveal characteristics common to many results in mathematics, and we would like to make three comments about these.

1. What at first might appear mysterious can often be made easy if you simply remember what you are talking about. Thus someone who has never before seem the symbol $a \mid b$ might look at the statement of Lemma 2.1 and mistakenly conclude that it concerned some highly advanced, sophisticated (and possibly frightening) fact. Yet when one remembers that "$a \mid b$" is simply a shorthand way of stating that "a divides b," then Lemma 2.1 is seen to be a quite easy and understandable fact. The moral of this story is that when faced with a new problem, first be sure that you understand what all the words and symbols mean. Sometimes this is easier said than done; nevertheless, when you have accomplished the deciphering, your task will be, if not easy, at least easier.

2. Do not lose sight of the forest for the trees. In the entire proof of Lemma 2.1 there is only one key idea and that is the use of the distributive law. The rest of the proof consists of making sure of what our goal is and what tools we are given with which to reach that goal. You must do this in any proof, whether or not you explicitly write out the details. Putting these ideas aside, you can see that it is the distributive law which really provides the result. As you read later proofs you should try to pick out the key idea or ideas. Soon you may find yourself measuring the sophistication of a proof by the number of key ideas it uses. Of course, as far as sophistication is concerned, Lemma 2.1 is but a country bumpkin, but as we said earlier, it is quite useful.

3. It is as important to understand what a result does *not* say as it is to understand what it does say. Lemma 2.1 says that if $u \mid v_1, u \mid v_2, \ldots, u \mid v_n$, then $u \mid (v_1 + v_2 + \cdots + v_n)$. It does not say that if $u \mid (v_1 + v_2 + \cdots + v_n)$, then $u \mid v_1, u \mid v_2, \ldots, u \mid v_n$. In fact, this last statement is blatantly false. For example, $6 \mid (7 + 3 + 2)$ but $6 \nmid 7$, $6 \nmid 3$, and $6 \nmid 2$.

Lemma 2.2. If u and v are integers and $u \mid v$, then, for any integer w, $u \mid vw$.

Proof. Since $u \mid v$, there is $c \in \mathbb{Z}$ such that $v = uc$. Then $vw = (uc)w = u(cw)$. Since $cw \in \mathbb{Z}$, by the definition $u \mid vw$. ∎

A corollary usually identifies a statement or a result which is an immediate logical consequence of another result. As an example:

Corollary 2.3. Let $u, v_1, v_2, \ldots, v_n$ be integers such that $u \mid v_1, u \mid v_2, \ldots, u \mid v_n$. Then for any choice of integers $w_1, w_2, \ldots, w_n$ we have that $u \mid (v_1 w_1 + v_2 w_2 + \cdots + v_n w_n)$.

Proof. By Lemma 2.2, $u \mid v_1 w_1, u \mid v_2 w_2, \ldots, u \mid v_n w_n$ and so by Lemma 2.1, $u \mid (v_1 w_1 + v_2 w_2 + \cdots + v_n w_n)$. ∎

Perhaps the key result on division is the theorem commonly referred to as the *Division Algorithm*.

Theorem 2.4. If $a, b \in \mathbb{Z}$ and $a > 0$, $b > 0$, then there exist integers q and r such that

$$a = bq + r \qquad \text{and} \qquad 0 \leq r < b$$

Before proving this, let us demonstrate that the theorem is a formalization of a fact you have known and used since grade school. If, for instance, $a = 317$ and $b = 5$, then dividing 317 by 5 gives 63 and a remainder of 2:

$$\begin{array}{r} 63 \\ 5\overline{)317} \\ 30 \\ \hline 17 \\ 15 \\ \hline 2 \end{array} = \text{remainder}$$

Thus $\frac{317}{5} = 63 + \frac{2}{5}$; multiplying by 5 we get $317 = 63(5) + 2$. The Division Algorithm in this case asserts the existence of $q = 63$ and $r = 2$. Notice that $0 \leq 2 < 5$. Thus what Theorem 2.4 says is that you can always perform the long division $a \div b$ and obtain a nonnegative remainder r which is less than the number you are dividing by, b. We now proceed to the proof of a fact that you may have taken for granted for a long time.

Proof.(of the Division Algorithm). Let us list the nonnegative multiples of b in increasing order:

$$0 = 0b < 1b < 2b < 3b < 4b < \cdots$$

Since $a > 0$ and fixed, and the numbers in the list get arbitrarily large, we may conclude that a lies between two successive multiples of b. We can make this statement precise using the Principle of Well-Ordering. Let $D = \{m \in \mathbb{Z} \mid m > 0 \text{ and } a < mb\}$. Then $D \neq \varnothing$ because $a + 1 \in D$, and D does not have arbitrarily small elements, since everything in the set D is positive. Thus D has a smallest element, which we call d. Its two distinguishing properties are that $d > 0$ and db is the first positive multiple of b to exceed a. Since $d - 1 \notin D$, we have

$$(d - 1)b \leq a$$

Our choices for q and r are then $q = d - 1$ and $r = a - (d - 1)b = a - qb$. Rewriting the inequality $(d - 1)b \leq a < db$ in terms of $q = d - 1$, we get $qb \leq a < (q + 1)b$. Since we have chosen $r = a - qb$, we solve for a and get $a = qb + r$. The only thing that remains is the verification that $0 \leq r < b$. However, $qb \leq a$, so $0 \leq a - qb = r$. On the other hand, $a < (q + 1)b = qb + b$, so substituting $qb + r$ for a yields $qb + r < qb + b$, and so $r < b$ by canceling qb from both sides. This completes the proof. ∎

A word about the proof is in order. Once we had chosen q and r via the Principle of Well-Ordering, we had to show that r satisfied the requirements of the statement and so we had to check two things: $0 \leq r$ and $r < b$. We also knew two things: $qb \leq a$ and $a < (q + 1)b$. Of course, $qb \leq a$ gave us $0 \leq r$ and $a < (q + 1)b$ gave us $r < b$. In more sophisticated proofs, the relationships between the various known facts and the desired goals may not be as clear. However, we encourage you to trace down those relationships which make the proof work. In this way both your understanding and appreciation of the arguments involved will be enhanced.

We have not quite told you everything about the Division Algorithm. For positive integers a and b we have shown that there are q and r in $\mathbb{Z}$ such that $a = bq + r$ and $0 \leq r < b$. We now point out there is only one such pair q, r which satisfies all the conditions of the theorem. For example, the equation

$$317 = 5q + r$$

has infinitely many pairs of solutions. For instance, possible choices for q and r are $q = 65$, $r = -8$ or $q = 43$, $r = 102$, among others. However, there is only one solution to the equation which satisfies the additional condition $0 \leq r < 5$ and that is the pair $q = 63$, $r = 2$. As an exercise, you are asked to show that if $a > 0$, $b > 0$, then there exists exactly one pair of numbers q, r with the properties that $a = bq + r$ *and* $0 \leq r < b$. (See Exercise 2.B.2.)

There is still more to be said regarding the Division Algorithm. Although Theorem 2.4 is stated in its most useful case, it is not really necessary to insist that both a and b be positive integers. For $a < 0$ and $b > 0$ nothing changes in the proof except that we consider negative multiples of b:

$$\cdots(-3)b < (-2)b < (-1)b < 0 \cdot b$$

For $a > 0$ and $b < 0$ we would consider negative multiples of b and require r to satisfy $0 \leq r < -b$. For example, if $a = 317$, $b = -5$, then $317 = (-5)(-63) + 2$ and $0 \leq 2 < -(-5)$. Finally, if $a < 0$ and $b < 0$, then we would consider positive multiples of b but require r to satisfy $0 \leq r < -b$.

As a parting thought we suggest that you look up the word "algorithm" if you are not familiar with its meaning. The algorithm referred to in the name of Theorem 2.4 is the process you learned sometime ago, long division.

2.2 Greatest Common Divisors

So far, perhaps, not much in this chapter has been very surprising. It is now time to begin to dig deeper into the properties of division.

Definition. Let a and b be integers; the integer d is called the *greatest common divisor* (GCD) of a and b if d satisfies the following two conditions:

(i) $d \mid a$ and $d \mid b$.

(ii) If $c \in \mathbb{Z}$ such that $c \mid a$ and $c \mid b$, then $c \leq d$.

Let us clarify the definition. Condition (i) says that d is an integer that divides both a and b. Thus d is a common divisor of both a and b. Condition (ii) says that d is greater than or equal to any other common divisor of a and b. Since we will be using greatest common divisors from now on, we introduce a notation for this concept.

Notation. The greatest common divisor of two integers a and b will be denoted by the symbol (a, b).

As examples you can verify easily that the following are true:

$$(6, 4) = 2, \qquad (12, 18) = 6, \qquad (-12, 18) = 6, \qquad (-5, 0) = 5$$

The definition of (a, b) gives the implication that there is only one greatest common divisor. That this is actually the case can be verified as follows. Suppose that d_1 is an integer which also satisfies conditions (i) and (ii). By (i), $d_1 \mid a$ and $d_1 \mid b$, hence $d_1 \leq d$, by using condition (ii) for d. Now interchange the roles of d_1 and d. By condition (i) $d \mid a$ and $d \mid b$, hence $d \leq d_1$, using condition (ii) for d_1. Hence $d_1 \leq d$ and $d \leq d_1$, so $d = d_1$. Another consequence of the definition is that for integers a and b, the greatest common divisor does not exist if both a and b are zero, since every integer divides zero. This raises the question: Given two integers a and b with at least one of a and b being nonzero, does the greatest common divisor exist? The answer is yes, since if $c \neq 0$, then there are only finitely many divisors of c (every divisor is between $-|c|$ and $|c|$). In particular, there can only be finitely many common divisors of a and b (there are some; 1 is always a common divisor). This finite set obviously has a greatest element.

Your only previous contact with greatest common divisors may have occurred when you studied reducing fractions to lowest terms. Let us assure you that their utility goes beyond reducing fractions or even as a source of homework problems for mathematics teachers. One of the uses that is accessible to us is in studying when the linear equation

$$aX + bY = c, \qquad a, b, c \in \mathbb{Z}$$

has solutions $x, y \in \mathbb{Z}$. We stress that we are seeking solutions which are integers. After trying many equations, such as $6X + 4Y = 14$, which has as solutions $x = 1$, $y = 2$ and $x = 3$, $y = -1$ as well as many others, or the equation $18X + 12Y = 8$, which apparently has no integer solutions (try it!), mathematicians probably began to suspect that $aX + bY = c$ would have a solution (or solutions) whenever (a, b) was a divisor of c. Thus in the first case $(6, 4) = 2$ and $2 \mid 14$, while in the second case $(18, 12) = 6$ and $6 \nmid 8$. This is in fact the case, as we will show; we will also discuss how the multiple solutions arise, such as for the equation $6X + 4Y = 14$. You should keep in mind that what we are about to present are the polished results, the outcome of many hours of work of people who perhaps looked at dozens of examples and tried many different techniques until a successful way was found. Although the facts are lost in history, it is more than likely that the problem of finding integer solutions to the equation $aX + bY = c$ was not devised as an application of greatest common divisors but, rather, that greatest common divisors were invented as a tool for solving equations such as $aX + bY = c$. We will return to this equation soon.

There are two different standard approaches to the development of the properties of greatest common divisors. We will present the longer of these two methods. Later we will outline the other method and discuss the relative merits of each approach. We first single out for consideration certain subsets of $\mathbb{Z}$.

Definition. An *ideal in* $\mathbb{Z}$ is a subset I of $\mathbb{Z}$ that has the following three properties:

(i) $I \neq \varnothing$, the empty set.

(ii) If $u \in I$ and $v \in I$, then $u + v \in I$.

(iii) If $u \in I$, then $ku \in I$ for all integers k.

EXAMPLE. A subset of $\mathbb{Z}$ that is an ideal in $\mathbb{Z}$ is the set I consisting of all even integers,

$$I = \{2k \mid k = 0, \pm 1, \pm 2, \ldots\}$$

This is so because (i) $I \neq \varnothing$, (ii) the sum of any two even integers is an even integer, and (iii) the product of an even integer and *any* integer is again an even integer. Another example is the set consisting of 0 alone, $J = \{0\}$. Of course $J \neq \varnothing$. To verify (ii) there is only one choice for u and v, namely $u = 0$, $v = 0$, and $u + v = 0 + 0 = 0 \in J$. For (iii) again there is only one choice for u, $u = 0$ and $k \cdot u = k \cdot 0 = 0 \in J$ for any integer k.

EXAMPLE. Not every nonempty subset of $\mathbb{Z}$ is an ideal in $\mathbb{Z}$; here are two subsets of $\mathbb{Z}$ which fail to be ideals. The set $P = \{k \in \mathbb{Z} \mid k > 0\}$ (the set of all positive integers) fails to satisfy condition (iii) because while $3 \in P$, $-1 \cdot 3 \notin P$. Notice that P satisfies conditions (i) and (ii). As a second example, the set D of all odd integers satisfies condition (i) but neither (ii) nor (iii): $3 \in D$, $-5 \in D$, but $3 + (-5) \notin D$ and $2 \cdot (-3) \notin D$. [Of course, multiplying 3 by an odd integer does yield another number of D, but condition (iii) requires that 3 multiplied by *every* integer belong to D, if D is going to be an ideal in $\mathbb{Z}$.]

Let us now provide you with a method for finding many ideals in $\mathbb{Z}$.

Lemma 2.5. If a is any integer, then the set

$$I = \{u \mid u = ka \text{ for some } k \in \mathbb{Z}\} = \{ka \mid k = 0, \pm 1, \pm 2, \ldots\}$$

is an ideal in $\mathbb{Z}$.

Proof. We must verify that I satisfies the three conditions of the definition. Before proceeding, however, let us examine I a little closer. The definition of I says that an integer u belongs to I if and only if $u = ka$ for some integer k. In terms of division we see that an integer u belongs to I if and only if $a \mid u$; thus I is precisely the set of all integers divisible by a. With this description of I and our previous lemmas, it is now a routine matter to verify conditions (i), (ii), and (iii). Certainly, $a \mid a$, so $a \in I$ and hence $I \neq \varnothing$, showing that (i) holds. If $u \in I$, $v \in I$, then $a \mid u$ and $a \mid v$, so by Lemma 2.1, $a \mid u + v$ and hence $u + v \in I$; thus (ii) holds. Finally, if $u \in I$ and $k \in \mathbb{Z}$, then $a \mid u$ and so by Lemma 2.2, $a \mid ku$; hence $ku \in I$, showing that (iii) holds. ∎

We now have a large number of ideals at our command, which, as the proof of Lemma 2.5 has shown, are somehow related to the process of division. Thus some ideals in $\mathbb{Z}$ are the sets

$$\{\ldots, -3, -2, -1, 0, 1, 2, 3, \ldots\} = \{k \cdot 1 \mid k \in \mathbb{Z}\}$$
$$\{\ldots, -4, -2, 0, 2, 4, \ldots\} = \{k \cdot 2 \mid k \in \mathbb{Z}\}$$
$$\{\ldots, -200, -100, 0, 100, 200, \ldots\} = \{k \cdot 100 \mid k \in \mathbb{Z}\}$$

Are there any other ideals in $\mathbb{Z}$ besides sets of this type? No!

Proposition 2.6. If I is any ideal in $\mathbb{Z}$, then there is an integer a such that $I = \{ka \mid k \in \mathbb{Z}\}$. Moreover a can be chosen to be nonnegative.

Proof. Since I is an ideal in $\mathbb{Z}$, condition (i) ensures that $I \neq \varnothing$, so let $x \in I$. Now $0 = 0 \cdot x$, so by condition (iii) $0 \in I$, and it might happen that 0 is the only integer in I. If this is the case, let $a = 0$, then $I = \{0\} = \{k \cdot 0 \mid k \in \mathbb{Z}\}$, and we are finished. Thus we can assume that, in addition to containing 0, I contains nonzero integers. If $x \in I$ and $x \neq 0$, then by condition (iii) $-x = (-1)x \in I$. Since one of x and $-x$ is positive, we now know that I contains positive integers. By the Principle of Well-Ordering (applied to the set $D = \{x \in \mathbb{Z} \mid x \in I \text{ and } x > 0\}$) there is a smallest positive integer which is contained in I, and this integer is our choice for a. That is, $a \in I$, $a > 0$ and any positive integer less than a does not belong to I. We must now show that $I = \{ka \mid k \in \mathbb{Z}\}$. First, by condition (iii), $ka \in I$ for any $k \in \mathbb{Z}$ since $a \in I$. Thus $\{ka \mid k \in \mathbb{Z}\} \subseteq I$. We must now show that $I \subseteq \{ka \mid k \in \mathbb{Z}\}$; that is, if $u \in I$, then $u = ka$ for some $k \in \mathbb{Z}$. Put another way we must show that if $u \in I$, then $a \mid u$. By the Division Algorithm we know that there are integers q, r such that

$$u = aq + r \qquad \text{with } 0 \leq r < a$$

If r happens to be 0, we have exactly what we need, $a \mid u$. So let us see what happens under the assumption that $r \neq 0$. In this case $r > 0$, $r < a$ and $r = u - qa$. However, $u \in I$ and $-qa = (-q)a \in I$ by (iii), so $r = u - qa = u + (-qa) \in I$ by condition (ii). Thus the assumption that $r \neq 0$ provides us with a positive integer less than a which belongs to I, namely r. This is contradictory to our choice of a as the least positive integer in I. We conclude that r must be zero, so $u = aq$ and hence $I \subseteq \{ka \mid k \in \mathbb{Z}\}$. ∎

We now know exactly what all ideals in $\mathbb{Z}$ look like; they are just those subsets of $\mathbb{Z}$ which consist of all multiples of some fixed integer a. At this point you might claim that we have led you on a wild goose chase. Why, you may ask, did we define an ideal as we did when we could easily have defined an ideal as any subset of $\mathbb{Z}$ of the form $\{ka \mid k \in \mathbb{Z}\}$, where a is fixed, since that is what ideals turn out to be anyway? One reason is that we now have two descriptions of an ideal, one complicated and one simple. Suppose, by way of example, that we consider $I = \{18x + 12y \mid x, y \in \mathbb{Z}\}$. It is not hard to see that this is an ideal. That (i) holds follows from letting $x = 0 = y$ so that $0 \in I$. Next, if $u \in I$, $v \in I$, then $u = 18x + 12y$ and $v = 18x_1 + 12y_1$ for some integers x, x_1, y, and y_1. Hence $u + v = 18(x + x_1) + 12(y + y_1) \in I$. Finally, for any integer k, $ku = k(18x + 12y) = 18(kx) + 12(ky) \in I$.

Now that we see that I is an ideal, we know that there is an integer a such that $I = \{ka \mid k \in \mathbb{Z}\}$. That is, since I satisfies the complicated description of an ideal, we know that it also satisfies the simple description and so is actually a simple set (and this was not obvious at the start).

Let us continue examining $I = \{18x + 12y \mid x, y \in \mathbb{Z}\}$ to get a preview of what is to come. We just saw that there is an a such that $I = \{ka \mid k \in \mathbb{Z}\}$. Since I contains $18(1) + 12(0) = 18$, $18(0) + 12(-1) = -12$, $18(1) + 12(-1) = 6$, $18(1) + 12(-2) = -6$, and so on, you might soon guess that $a = 6$. This is correct, as we shall soon prove [the key fact will be that $(18, 12) = 6$]. Now we have $I = \{18x + 12y \mid x, y \in \mathbb{Z}\} = \{k \cdot 6 \mid k \in \mathbb{Z}\}$. Earlier we considered the equation $18X + 12Y = 8$ and suspected that it had no solutions. Now we can see why not. For any integers x and y, $18x + 12y \in I = \{k6 \mid k \in \mathbb{Z}\}$, so $18x + 12y$ must be a multiple of 6. However, 8 is not a multiple of 6, so $18x + 12y = 8$ is impossible. On the other hand, if we had considered $18X + 12Y = 54$, then since $6 \mid 54$, we have $54 \in \{k6 \mid k \in \mathbb{Z}\} = I = \{18x + 12y \mid x, y \in \mathbb{Z}\}$, and thus for some $x \in \mathbb{Z}$, $y \in \mathbb{Z}$ we will have $18x + 12y = 54$. As we said, this has been a preview. We now plunge into the details.

Proposition 2.7. Let a and b be in $\mathbb{Z}$. Then

$$I = \{ax + by \mid x, y \in \mathbb{Z}\}$$

is an ideal in $\mathbb{Z}$. If not both of a and b are zero, then

$$I = \{kd \mid k \in \mathbb{Z}\} \qquad \text{where } d = (a, b)$$

Proof. We leave it to you as an exercise to verify that I is an ideal in $\mathbb{Z}$; the proof is similar to that constructed for $\{18x + 12y \mid x, y \in \mathbb{Z}\}$. Knowing that I is an ideal, from Proposition 2.6 we conclude that there is an integer $d \geq 0$

such that $I = \{kd \mid k \in \mathbb{Z}\}$. If both a and b are zero, then $I = \{0\}$ and $d = 0$ in this case. Thus we may assume that $a \neq 0$. Then we know that $d > 0$. We will now show that $d = (a, b)$, and to do this we have to verify two things. First we must show that $d \mid a$ and $d \mid b$. Now $a = a \cdot 1 + b \cdot 0$, so that $a \in I$. However, $I = \{kd \mid k \in \mathbb{Z}\}$, so $a = td$ for some t. This says that $d \mid a$, and a parallel argument shows that $d \mid b$ also. We next show that any common divisor of a and b is no larger than d. Suppose that $c \mid a$ and $c \mid b$; then $a = cm$, $b = cn$ for some integers m and n. The fact that $I = \{kd \mid k \in \mathbb{Z}\}$ says, in particular, that $d = 1 \cdot d \in I$ (condition (ii)). The original description of I was $I = \{ax + by \mid x, y \in \mathbb{Z}\}$; hence since $d \in I$, then there are integers x_0 and y_0 such that $ax_0 + by_0 = d$. Replacing a by cm and b by cn, we get

$$d = ax_0 + by_0 = (cm)x_0 + (cn)y_0 = c(mx_0 + ny_0)$$

Thus $c \mid d$. Since $d > 0$, we easily see that $c \leq d$ [Exercise 1.A.1(e)]. ∎

A fairly obvious but very important consequence of this proposition is worth stating as a theorem.

Theorem 2.8. Let a and b be integers, not both zero. If $d = (a, b)$, then there exist integers x_0 and y_0 such that $d = ax_0 + by_0$.

Proof. By Proposition 2.7, $\{ax + by \mid x, y \in \mathbb{Z}\} = \{kd \mid k \in \mathbb{Z}\}$; hence $d = ax_0 + by_0$ for some integers x_0 and y_0. ∎

2.3 The Equation $aX + bY = c$

As we promised, greatest common divisors can be useful in studying the equation $aX + bY = c$, where a and b are integers, not both zero. This particular equation is known as a *linear* Diophantine equation. Apparently, it and similar equations were first studied by the Greek mathematician Diophantus, who lived sometime during the third century A.D. The equation is a small first step into a branch of number theory known as Diophantine analysis. Other ventures into this fascinating area will take place in Chapter 7.

Proposition 2.9. Let a and b be integers, not both zero, and let $d = (a, b)$. The equation $aX + bY = c$ has integer solutions if and only if $d \mid c$.

Proof. In the statement the phrase "if and only if" occurs. A statement of the form "A if and only if B" is equivalent to having two statements: "If A, then B" and "If B, then A." Thus we must prove two things:

1. If the equation $aX + bY = c$ has integer solutions, then $d \mid c$.
2. If $d \mid c$, then the equation $aX + bY = c$ has integer solutions.

Let us consider the first statement. Suppose that the equation $aX + bY = c$ has integer solutions, say $x = x_0$, $y = y_0$. Then $c = ax_0 + by_0$. Since $d = (a, b)$ we have $d \mid a$ and $d \mid b$; hence by Corollary 2.3, $d \mid ax_0 + by_0$, that is, $d \mid c$. This takes care of statement 1. For statement 2, we assume that $d \mid c$. Thus $c \in \{kd \mid k \in \mathbb{Z}\}$. By Proposition 2.7, $\{kd \mid k \in \mathbb{Z}\} = \{ax + by \mid x, y \in \mathbb{Z}\}$, so $c \in \{ax + by \mid x, y \in \mathbb{Z}\}$. Another way of saying this is that there are integers x_0, y_0 such that $c = ax_0 + by_0$. ∎

While the previous proposition is satisfying in some respects, it and its proof do not tell us how to find solutions, although it does start us in the proper direction. If confronted with the equation

$$aX + bY = c, \qquad \text{not both } a, b \text{ zero}$$

we can first test to see if solutions exist. This is accomplished by first finding $d = (a, b)$ and seeing whether or not $d \mid c$. Let us suppose that $d \mid c$. Another proof of statement 2 that uses Theorem 2.8 and is more instructive is as follows. By 2.8, $d = ax_0 + by_0$ for some integers x_0 and y_0. Since $d \mid c$, $c = kd$ for some $k \in \mathbb{Z}$. Multiplying $ax_0 + by_0$ by k gives us $c = kd = k(ax_0 + by_0) = a(kx_0) + b(ky_0)$, so $x = kx_0, y = ky_0$ is a solution of the equation. This proof suggests, then, that the problem of solving the equation $aX + bY = c$ really reduces to finding integers x_0 and y_0 such that $(a, b) = d = ax_0 + by_0$, as predicted in Theorem 2.8. Before tackling this problem let us look at an equation with small numbers which is solvable.

Example. For the equation $6X + 4Y = 14$, we have $(6, 4) = 2$ and $14 = 7 \cdot 2$. In this case it is easy to see that $2 = 6 \cdot 1 + 4(-1)$, so multiplying by 7 gives us $14 = 6(7) + 4(-7)$. Thus $x = 7, y = -7$ is a solution for the equation. But there are other solutions, such as $x = 1$, $y = 2$ or $x = 3$, $y = -1$ or $x = -1, y = 5$.

The previous example suggests that the question seems to be not only if there are any solutions, but also how many there are, and how we find all of them. Rather than simply giving you the answers, we think it will be more beneficial if we can show you how to find them. The key to this task, and many like it in mathematics, is to consider many examples. We have seen that for $x = 1$, 3, or 7, there is a corresponding y which is, respectively, $y = 2, -1, -7$. What about $x = 5$? Well, $14 = 6 \cdot 5 + 4 \cdot y$ yields $y = \dfrac{14 - 30}{4} = -4$. Thus we might suspect that odd values of x provide solutions.

Let us test this one more time with $x = 9$. Here $14 = 6 \cdot 9 + 4y$, so $y = \dfrac{14 - 6 \cdot 9}{4} = -10$. In order to see if there is any pattern to these, let us arrange the solutions obtained so far in some nice manner. For example, we can list them in increasing order of x:

$$x = -1, y = 5; \quad x = 1, y = 2; \quad x = 3, y = -1; \quad x = 5, y = -4;$$
$$x = 7, y = -7; \quad x = 9, y = -10$$

We see that the x values are increasing by 2, while the y values are decreasing by 3. If this pattern is indeed what we are seeking, then we should be able to predict solution pairs. The next solution in our ordering should be $x = 11$, $y = -13$, while the solution preceding $x = -1, y = 5$ should be $x = -3$, $y = 8$. A quick verification shows that this is indeed the case. For this equation it appears that there will be infinitely many solutions and that they follow a nice pattern.

Trying another equation, $9X + 15Y = 6$, we see that $3 = (9, 15)$ and $3 \mid 6$, so solutions exist. Searching yields the following pairs of solutions:

$$x = -6, y = 4; \quad x = -1, y = 1; \quad x = 4, y = -2;$$
$$x = 9, y = -5$$

Here the x values increase by 5 while the y values decrease by 3. With the equation $25X + 30Y = 15$, we find that $5 = (25, 30)$, $5 \mid 15$ and that the pairs

$$x = -3, y = 3; \quad x = 3, y = -2; \quad x = 9, y = -7$$

are solutions. Here x increases by 6 while y decreases by 5.

In each case it appears that there will be an infinite number of pairs of solution and that, in fact, there is a nice pattern which describes these solutions. Now, can we deduce the pattern? With $6X + 4Y = 14$, x increases by 2 while y decreases by 3. With $9X + 15Y = 6$, x increases by 5 while y decreases by 3. With $25X + 30Y = 15$, x increases by 6 while y decreases by 5. If you do not yet see what is happening, try some more examples and you will eventually notice that $(6, 4) = 2$ and $4/2 = 2$, while $6/2 = 3$. Also, $(9, 15) = 3$ and $15/3 = 5$, while $9/3 = 3$. Finally, $(25, 30) = 5$, while $30/5 = 6$ and $25/5 = 5$. If this observation is indeed the key, then, after examining the equation $8X + 12Y = 4$ and finding the solution $x = -1, y = 1$, it should be easy to predict other solutions. Since $(8, 12) = 4$, $12/4 = 3$ and $8/4 = 2$, we should expect the x values to increase by 3 as the y values decrease by 2 (or decrease the x values by 3 as the y values increase by 2). We quickly verify that

$$x = -1 + 3, y = 1 - 2; \qquad x = -1 + 3 \cdot 2, y = 1 - 2 \cdot 2;$$
$$x = -1 + 3 \cdot 3, y = 1 - 2 \cdot 3$$

are all solutions.

Having made a guess in this case we would like to state this guess in a manner that will work for all equations and then prove the statement. We should warn you that there are instances where this process of deducing general statements from examples can lead to statements that may be quite difficult to prove. In fact, as we will see later, there are some statements (called *conjectures*) which have been based on the behavior of many, many examples and which have not yet been proven (or shown to be false). Fortunately, our guess is not in this category.

Proposition 2.10. Let a and b be integers, not both zero, and let $d = (a, b)$. The equation $aX + bY = c$ has integer solutions if and only if $d \mid c$. If $x = r, y = s$ is a solution of $aX + bY = c$, and if $a = du, b = dv$, then for any $k \in \mathbb{Z}$,

$$x = r + vk, \quad y = s - uk$$

is also a solution of $aX + bY = c$.

Proof. We have already shown in Proposition 2.9 that $aX + bY = c$ has solutions if and only if $d \mid c$. If $x = r, y = s$ is a solution of $aX + bY = c$, we have that $ar + bs = c$. In addition,

$$\begin{aligned} a(r + vk) + b(s - uk) &= ar + a(vk) + bs - b(uk) \\ &= (ar + bs) + (du)(vk) - (dv)(uk) = c + 0 = c \end{aligned}$$

and we are done. ∎

As Proposition 2.10 tells us, once we have one solution we can find many other solutions of the equation $aX + bY = c$. What remains to be done is to show that the process actually accounts for all solution pairs. That is, we must show that if x_1, y_1 is a solution, then (using the notation of proposition 2.10) $x_1 = r + vk$ and $y_1 = s - uk$ for some integer k. While the proof of this fact is not too difficult, it does require a bit of preparation.

Lemma 2.11. Let $d = (a, b)$ and suppose that $a = de, b = df$. Then $(e, f) = 1$.

Proof. Let $c = (e, f)$. Then we may write $e = ch, f = cg$ for some integers h, g. Thus $a = de = (dc)h$ and $b = df = (dc)g$. Therefore, dc is a common divisor of a and b and since $d = (a, b)$, we have $dc \leq d$ and so $c \leq 1$. Recalling that greatest common divisors are always positive, we conclude that $c = 1$. ∎

The situation where two integers have greatest common divisor 1 is, as we shall see, important enough to merit an appropriate appellation.

Definition. Two integers u and v are *relatively prime* if and only if $(u, v) = 1$.

It will become clear after reading Chapter 3 why the term "relatively prime" is used to designate this particular situation.

We should point out that an integer u can be a divisor of a product fh without being a divisor of either f or h (Exercise 2.A.2). However, if u is relatively prime to one of the factors, then we can conclude something worthwhile.

Lemma 2.12. Let u, f, and h be integers such that $u \mid fh$. If $(u, f) = 1$, then $u \mid h$.

Proof. Using Theorem 2.8, we know that there are integers x and y such that $ux + fy = 1$. Multiplying by h, we get $u(hx) + (fh)y = h$. Now $u \mid fh$ and of course $u \mid u$, so we conclude by Corollary 2.3 that $u \mid h$. ∎

Let us next consider two relatively prime integers, u and v. If for some $g, h \in \mathbb{Z}$ we have $ug = vh$, then Lemma 2.12 assures us that $u \mid h$ and, at the same time, that $v \mid g$. Our next lemma sharpens this observation.

Lemma 2.13. Suppose that $u, v \in \mathbb{Z}$ with $(u, v) = 1$. If $g, h \in \mathbb{Z}$ such that $ug = vh$, then there exists $t \in \mathbb{Z}$ such that $h = ut$ and $g = vt$.

Proof. As we noted prior to the statement of the lemma, $u \mid h$, so $h = ut$ for some $t \in \mathbb{Z}$. Then $ug = vh = v(ut) = u(vt)$, so $g = vt$. ∎

Returning to the equation $aX + bY = c$ we summarize what we have.

Theorem 2.14. Let $a, b \in \mathbb{Z}$ not both zero and let $d = (a, b)$.

(i) The equation $aX + bY = c$ has solutions if and only if $d \mid c$.

(ii) If $x = r$, $y = s$ is a solution of $aX + bY = c$ and if $a = du$, $b = dv$ then for any $k \in \mathbb{Z}$, $x = r + vk$, $y = s - uk$ is also a solution of the equation $ax + by = c$. Conversely, if x_1, y_1 is any solution of $aX + bY = c$, then there is an integer k such that $x_1 = r + vk$, $y_1 = s - uk$.

Proof. Part (i) as well as the first sentence of (ii) has already been established. Thus suppose that we have at hand one solution $x = r$, $y = s$ of the equation $aX + bY = c$, and suppose that $x = x_1, y = y_1$ is any other solution. We then have

$$ar + bs = c = ax_1 + by_1$$

That is, $a(r - x_1) = b(y_1 - s)$. Since $a = du$ and $b = dv$, substituting these expressions and canceling d, we get $u(r - x_1) = v(y_1 - s)$. By Lemma 2.11, $(u, v) = 1$; hence applying Lemma 2.13 with $g = r - x_1$ and $h = y_1 - s$, we know that there exists $t \in \mathbb{Z}$ such that $y_1 - s = ut$ and $r - x_1 = vt$. Thus $y_1 = s + ut$ and $x_1 = r - vt$. Finally, letting $k = -t$, we have $x_1 = r + vk$, $y_1 = s - uk$. ∎

This theorem, while appearing to be conclusive, has a glaring deficiency. Its application requires knowing at least one solution to the equation $aX + bY = c$, where $d = (a, b)$ and $d \mid c$. Let $a = du$, $b = dv$ and $c = dw$. Recalling the discussion following Proposition 2.9, the problem of solving $aX + bY = c$ is really that of finding integers x_0 and y_0 such that $ax_0 + by_0 = d$. For if we can find such x_0, y_0, then multiplying by w gives $a(x_0w) + b(y_0w) = c$; hence $r = x_0w$, $s = y_0w$ is a solution. Is there a method, other than trial and error, for finding such x_0 and y_0? The answer is yes, and the method is known as *Euclid's Algorithm*. It depends on Theorem 2.4 and the following fact.

Lemma 2.15. Let $a, b \in \mathbb{Z}$. If $a = bq + r$, then $(a, b) = (b, r)$.

Proof. Suppose that $d = (a, b)$ and $e = (b, r)$. Since $d \mid a$, $d \mid b$, and $r = a - bq$, Corollary 2.3 tells us that $d \mid r$. However, $e = (b, r)$, so $d \leq e$ since d is a common divisor of b and r. On the other hand, $e \mid b$ and $e \mid r$, so $e \mid a$ because $a = bq + r$. Thus $e \leq d$. We conclude that $e = d$. ∎

EXAMPLE. Let us use the Division Algorithm (Theorem 2.4) together with Lemma 2.15 to calculate (8216, 1508). First,

$$8216 = 5(1508) + 676$$

Hence (8216, 1508) = (1508, 676). Next,

$$1508 = 2(676) + 156$$

so (1508), 676) = (676, 156). Continuing, we obtain

$$676 = 4(156) + 52 \quad \text{and} \quad (676, 156) = (156, 52)$$
$$156 = 3(52) + 0 \quad \text{and} \quad (156, 52) = (52, 0)$$

However, (52, 0) = 52, so stringing together the equalities, we have 52 = (52, 0) = (156, 52) = (676, 156) = (1508, 676) = (8216, 1508). We glean even more from the process. For we notice that from the equality $676 = 4(156) + 52$, we get

$$52 = (676)(1) + (156)(-4)$$

so 52 has been expressed as a combination of 676 and 156. Going one step further, we see that from the equality $1508 = 2(676) + 156$ we have

$$156 = 1508 + (-2)(676)$$

If we substitute this value for 156 in the expression for 52, we obtain

$$\begin{aligned} 52 &= (676)(1) + (156)(-4) = (676)(1) + [1508 + (-2)(676)](-4) \\ &= (676)(9) + (1508)(-4) \end{aligned}$$

Thus 52 has been expressed as a combination of 676 and 1508. Notice that we have solved the equations

$$676X + 156Y = 52 \qquad \text{and} \qquad 676X + 1508Y = 52$$

and that $52 = (676, 156)$ and $52 = (676, 1508)$. Of course, we would like to solve $8216X + 1508Y = 52$, and our two steps have brought us closer to this. Pushing further, from the first equality, $8216 = 5(1508) + 676$, we get $676 = 8216 + (-5)(1508)$. Substituting in $52 = (676)(9) + (1508)(-4)$ yields

$$\begin{aligned} 52 &= [8216 + (-5)(1508)](9) + (1508)(-4) \\ &= (8216)(9) + (1508)(-49) \end{aligned}$$

Hence we have accomplished what Theorem 2.8 predicted; we have found integers $x_0 (x_0 = 9)$ and $y_0 (y_0 = -49)$ such that

$$(8216, 1508) = 52 = 8216x_0 + 1508y_0$$

We will rewrite the steps so that you can see how this straightforward process works:

$$\begin{aligned} 8216 &= 5(1508) + 676 \\ 1508 &= 2(676) + 156 \\ 676 &= 4(156) + 52 \\ 156 &= 3(52) + 0 \end{aligned}$$

Then, working with the remainders,

$$\begin{aligned} 52 &= 676 + (-4)(156) \\ \underline{156} &= 1508 + (-2)(676) \\ \underline{676} &= 8216 + (-5)(1508) \end{aligned}$$

Thus

$$\begin{aligned}52 &= 676 + (-4)(156) \\ &= 676 + (-4)[1508 + (-2)(676)] \\ 52 &= 676(9) + (1508)(-4) \\ &= [8216 + (-5)(1508)](9) + (1508)(-4) \\ 52 &= (8216)(9) + (1508)(-49)\end{aligned}$$

We have underlined the numbers that are substituted *successively* rather than simultaneously.

EXAMPLE. Having performed the previous calculations, let us put them to use to solve the Diophantine equation

$$8216X + 1508Y = 260$$

We first calculate $(8216, 1508) = 52$, then check that $260 = (52)5$. Thus $52 \mid 260$; hence solutions exist. Next we express 52 as a combination of 8216 and 1508:

$$(8216)(9) + (1508)(-49) = 52$$

Multiplying by 5 gives

$$(8216)(45) + (1508)(-245) = 260$$

so one solution to the equation is $r = 45$, $s = -245$. To complete the problem, we describe all other solutions. In the notation of Theorem 2.14, we have $u = 158$ and $v = 29$, since $a = 8216 = (52)(158)$ and $b = 1508 = (52)(29)$. Therefore, all solutions are given by

$$x = 45 + 29k, \qquad y = -245 - 158k$$

where $k = 0, \pm 1, \pm 2, \ldots$.

EXAMPLE. An alternative step that you may wish to pursue in solving $8216X + 1508Y = 260$ is to divide by $52 = (8216, 1508)$. That is reduce the equation to $158X + 29Y = 5$. In Exercise 2.A.21 you are asked to show that division by the greatest common divisor does not change the solution set. Now proceed as before:

$$\begin{aligned}&\text{(i)} & 158 &= 5(29) + 13 \\ &\text{(ii)} & 29 &= 2(13) + 3 \\ &\text{(iii)} & 13 &= 4(3) + 1 \\ && 3 &= 3(1) + 0\end{aligned}$$

Hence $1 = (158, 29)$, a fact predicted by Lemma 2.11. Working backward, we have from lines (iii), (ii), and (i),

$$1 = 13 + (-4)3, \qquad 3 = 29 + (-2)13, \qquad 13 = 158 + (-5)29$$

Hence

$$\begin{aligned} 1 &= 13 + (-4)[29 + (-2)13] = (9)(13) + (-4)(29) \\ &= (9)[158 + (-5)29] + (-4)(29) = (158)(9) + (29)(-49) \end{aligned}$$

That is,

$$1 = (158)(9) + (29)(-45)$$

Multiplying by 5 yields

$$158(45) + 29(-245) = 5$$

Hence $r = 45$, $s = -245$ is a solution to $158X + 29Y = 5$. Notice that this was the same solution that we obtained for $8216X + 1508Y = 260$. Since $(158, 29) = 1$ and $158 = 1 \cdot 158$, $29 = 1 \cdot 29$, we have $u = 158$, $v = 29$, so all solutions to $158X + 29Y = 5$ (and hence to $8216X + 1508Y = 260$) are given by $x = 45 + 29k$, $y = -245 - 158k$, where $k \in \mathbb{Z}$.

We conclude this chapter by providing you with a formal statement of Euclid's Algorithm and some remarks.

Theorem 2.16. Let $a, b \in \mathbb{Z}$ with $b > 0$. Then (a, b) is the last nonzero remainder r_k in the series of equations obtained from the Division Algorithm:

$$\begin{aligned} a &= bq_1 + r_1, & 0 < r_1 < b \\ b &= r_1 q_2 + r_2, & 0 < r_2 < r_1 \\ r_1 &= r_2 q_3 + r_3, & 0 < r_3 < r_2 \\ &\vdots & \vdots \\ r_{k-2} &= r_{k-1} q_k + r_k, & 0 < r_k < r_{k-1} \\ r_{k-1} &= r_k q_{k+1} + 0 \end{aligned}$$

The proof of this result is left as an exercise.

We mentioned earlier that there were two approaches to the study of GCDs. We have chosen to introduce ideals as our approach. The other standard technique is via Theorem 2.16, Euclid's Algorithm. In any case, however, the key result will be Theorem 2.8, asserting the existence of x_0 and y_0 such that

$$ax_0 + by_0 = (a, b)$$

The advantage of Euclid's Algorithm is that it provides a method for finding x_0 and y_0. On the other hand, the use of ideals is a good illustration of how a concept such as ideals, which on the surface may appear to be quite theoretical, can in fact have very concrete applications. In addition, the concept of ideals has been a very powerful and unifying tool in other branches of mathematics, and this approach emphasizes that unity. Finally, if we wish to study the equation $aX + bY + cZ = d$ or even, say, $a_1X_1 + a_2X_2 + a_3X_3 + \cdots + a_nX_n = b$ where $n \geq 3$, the ideal-theoretic approach works completely unchanged (see Exercise 4.B.13). The Euclidean Algorithm can still be used but requires suitable reworking.

EXERCISES 2.A

1. Find all positive integers k such that $k \mid 12$; $k \mid 25$; $k \mid 300$.
2. Find integers u, f, and h such that $u \mid fh$ but $u \nmid f$ and $u \nmid h$.
3. True or false: If $u \mid (f + h)$, then either $u \mid f$ or $u \mid h$. (If true, prove it; if false, provide a numerical example.)
4. True or false: If $b \mid c$ and $a \mid c$, then $a \mid b$. (If true, prove it; if false, provide a numerical example.)
5. Prove: If $a \mid b$ and $b \mid c$, then $a \mid c$.
6. Suppose that $a \mid (k - 1)$; show that $a \mid (k^2 - 1)$. Does $a \mid (k^3 - 1)$? Is there a general result?
7. Show that for all $a \in \mathbb{Z}$, $\dfrac{a^2 - a}{2} \in Z$.
8. Show that for all $a \in \mathbb{Z}$, $\dfrac{a^3 - a}{3} \in Z$.
9. Show that if $k^2 - 1$ is even, then either $k - 1$ or $k + 1$ is even.
10. Show that if $a \mid 2x - 3y$ and $a \mid 4x - 5y$, then $a \mid y$.
11. For the following pairs of integers u, v, find $d = (u, v)$ and express d in the form $ux + vy$ for suitable integers x and y:

    (a) (273, 1124) (b) (17, 141) (c) (25, -113)

12. Show that for all $k \in \mathbb{Z}$, $(k, k + 1) = 1$.
13. Let $a > 0$. Show that $(a, b) = a$ if and only if $a \mid b$.
14. Show that $(5x + 6, 5x + 8) = 1$ for all odd integers x.
15. Let $a, b \in \mathbb{Z}$. Show that if there exist integers u, v such that $au + bv = 1$, then $(a, b) = 1$.
16. Let I and J be ideals in $\mathbb{Z}$ and let $I + J = \{x + y \mid x \in I \text{ and } y \in J\}$.
    (a) Show that $I + J$ is an ideal in $\mathbb{Z}$.
    (b) By Proposition 2.6, there exist integers a, b, and c such that

    $$I = \{ka \mid k \in \mathbb{Z}\}, \qquad J = \{kb \mid k \in \mathbb{Z}\}, \qquad I + J = \{kc \mid k \in \mathbb{Z}\}$$

    How is c related to a and b? (*Hint:* Try some explicit values for a and b.)

17. Let $a_1, \ldots, a_n \in \mathbb{Z}$. Show that the set $I = \{a_1x_1 + \cdots + a_nx_n \mid x_1, \ldots, x_n \in \mathbb{Z}\}$ is an ideal in $\mathbb{Z}$.
18. Find a subset of $\mathbb{Z}$ that satisfies conditions (i) and (iii) in the definition of ideal but fails to satisfy condition (ii).
19. Write out formally the statement of Theorem 2.4 as it would appear if you were dealing with the case $a < 0, b < 0$.
20. Solve the following equations in integers. (If no solutions exist, tell why.)
    (a) $21X + 12Y = 36$ (b) $21X + 12Y = 55$
    (c) $17X + 141Y = 1$ (d) $15X + 21Y = 18$
    (e) $10X - 25Y = 37$ (f) $42X + 49Y = -21$
    (g) $222X + 174Y = 60$ (h) $52X + 143Y = 39$
21. Consider $aX + bY = c$. Let $d = (a, b)$ and suppose that $d \mid c$. Suppose that $a = da_1$, $b = db_1$, and $c = dc_1$. Show that $aX + bY = c$ and $a_1X + b_1Y = c_1$ have exactly the same solutions.
22. Assume that $d \mid n$. Show that $dc \mid n$ if and only if $c \mid (n/d)$.
23. Show that if $a \mid x$, $b \mid x$, and $(a, b) = 1$, then $ab \mid x$. (*Hint:* $1 = au + bv$, so multiply by x.)

EXERCISES 2.B

1. Assume that $0 \leq r < b$ and $0 \leq s < b$. Show that if $b \mid r - s$, then $r = s$.
2. In Theorem 2.4 show that the integers q and r are unique; that is, if p, s satisfy $a = bp + s$ and $0 \leq s < b$, then $p = q$ and $r = s$.
3. Let a, b, and c be integers with $c > 0$. Show that $c \mid a - b$ if and only if a and b have the same remainder after division by c.
4. Let $a, b, c \in \mathbb{Z}$ with $c > 0$.
    (a) Show that $(ca, cb) = c(a, b)$ by using the definition of GCD.
    (b) Use Proposition 2.7 to show that $(ca, cb) = c(a, b)$.
5. Derive Lemma 2.11 from the result obtained in Exercise 2.B.4.
6. Let $a, b \in \mathbb{Z}$; an integer n is called a *common multiple* of a and b if $a \mid n$ and $b \mid n$. A positive integer m is called a *least common multiple* of a and b provided that
    (i) m is a common multiple of a and b.
    (ii) If $n > 0$ is any common multiple of a and b, then $m \leq n$.
    (a) Show that a least common multiple exists for any two nonzero integers and that it is unique.
    (b) For nonzero a, b, let $[a, b]$ denote the least common multiple of a and b. Show that if $a > 0, b > 0$ then $[a, b] = ab/(a, b)$.
    (c) If $(a, b) = 1$, then $[a, b] = ab$.

7. Consider an alternative way of defining the GCD of two integers a, b not both zero: An integer $d > 0$ is a *greatest common divisor* of a and b provided that

(i) d is a common divisor of a and b.

(ii) If c is any common divisor of a and b, then $c \mid d$.

Show that this is equivalent to the definition given in the text and discuss the relative merits of each. For instance, from which of these definitions is it easier to conclude that GCDs exist? (*Hint:* To show that the two definitions are equivalent, use Theorem 2.8.)

8. Prove that $(a^5 - a)/5 \in \mathbb{Z}$ for any $a \in \mathbb{Z}$.

9. Prove that if $(a, c) = 1$ and $(b, c) = 1$, then $(ab, c) = 1$. (*Hint:* Write $1 = ax_0 + cy_0$ and $1 = bx_1 + cy_1$. Multiply these two equations together.)

10. Show that the following is equivalent to the definition given in the text: $I \subset \mathbb{Z}$ is an ideal in $\mathbb{Z}$ if (i) $I \neq \varnothing$; (ii) if $u \in I$ and $v \in I$, then $u + v \in I$; and (iii) if $u \in I$, then $-u \in I$.

11. In the Cartesian plane, graph $6X + 4Y = 14$. Discuss how the integral solutions to this can be found from the graph. We saw that when appropriately listed, the solution pairs have x increasing by 2 while y decreased by 3. What connection does this have to the slope of the graph? You should also see why these equations are called linear.

12. Suppose that we use $(a, b)\mathbb{Z}$ to denote the ideal $\{ax + by \mid x, y \in \mathbb{Z}\}$ and $d\mathbb{Z}$ to denote the ideal $\{kd \mid k \in \mathbb{Z}\}$. If $d = (a, b)$, discuss the fact that on the surface, $(a, b)\mathbb{Z}$ might be interpreted in two different ways, but that in fact both interpretations yield the same ideal.

13. Outline the various steps and major results one would need in solving an equation of the form $aX + bY + cZ = d$.

14. Compare Exercises 2.A.7, 2.A.8, and 2.B.8. Guess a general result.

3
Prime Numbers

CHAPTER 2 was devoted to examination of division of one integer by another integer. Given integers a and b, it may or may not happen that $a \mid b$. Of course, if $a = \pm 1$ or if $a = \pm b$, then a automatically divides b. It is a fact, moreover, that for some choices of b these will be the only possibilities for having b divisible by a. For example, if $b = 17$, then the only divisors of b are $1, -1, 17$, and -17. Numbers such as 17 play a central role in what is to follow and they will command our attention during this chapter.

3.1 Primes

We begin with a definition.

Definition. An integer $p > 1$ is a *prime number* if the only positive divisors of p are 1 and p.

Clearly 2, 3, 5, 7, 11 (and 17) are all prime numbers, while 15 is not a prime number since it is divisible by 3 as well as 1 and 15. A moment's thought will convince you that 2 is the only even prime number.

The reason that prime numbers are so important is that every integer greater than 1 is a product of prime numbers. We illustrate this assertion with the number 420. We have $420 = 42 \cdot 10$ and $42 = 7 \cdot 6$ while $10 = 2 \cdot 5$, so that $420 = 7 \cdot 6 \cdot 2 \cdot 5$. Now each of 7, 2, and 5 is a prime number. However, $6 = 2 \cdot 3$, giving us $420 = 7 \cdot 2 \cdot 3 \cdot 2 \cdot 5$, and now each factor is a prime number and 420 has been shown to be a product of primes. To prove the assertion that every integer greater than 1 is a product of prime numbers, we will use induction. A word of caution: the phrase "a product of prime numbers" allows the possibility of only one term occurring in a product; thus a prime number is already "a product of prime numbers."

Theorem 3.1. Every integer greater than 1 can be expressed as a product of prime numbers.

Proof. Let $S = \{n \in \mathbb{Z} \mid n > 1$ and n can be expressed as a product of primes$\}$. We wish to show that S consists of all integers greater than or equal to 2. By the Principle of Induction it is enough to show two things: (i) $2 \in S$, and (ii) for any $x \geq 2$, if $\{n \in \mathbb{Z} \mid 2 \leq n \leq x\} \subseteq S$, then $x + 1 \in S$. Clearly, (i) is satisfied since 2 is a prime number (and so automatically a product of prime numbers). Now we turn our attention to condition (ii). Suppose that $x > 2$ and that $\{n \in \mathbb{Z} \mid 2 \leq n \leq x\} \subseteq S$. We must show that $x + 1 \in S$. If $x + 1$ happens to be a prime number, then automatically $x + 1 \in S$ and we are finished. If $x + 1$ is not a prime number, then there is a positive integer a which divides $x + 1$ and which differs from both 1 and $x + 1$. Thus we have $a \mid (x + 1)$ and $2 \leq a \leq x$. Let $b = (x + 1)/a$. Then $x + 1 = ab$, so that $b \mid (x + 1)$. Since $a > 1$, $b < x + 1$, and since $a < x + 1$, $b > 1$. Therefore, $2 \leq b \leq x$. Thus a and b are both members of the set $\{n \in \mathbb{Z} \mid 2 \leq n \leq x\}$ and since we are assuming that this set is a subset of S, we conclude that both a and b belong to the set S. However, everything in S can be expressed as a product of prime numbers. Doing this for a and b, we have $a = p_1 p_2 \cdots p_k$ and $b = q_1 q_2 \cdots q_l$, where each p_i and each q_j is a prime number. Then $x + 1 = ab = (p_1 p_2 \cdots p_k)(q_1 q_2 \cdots q_l)$. Hence $x + 1$ is a product of prime numbers and we are done. ∎

We have already told you that 2, 3, 5, 7, 11, and 17 are prime numbers and you may have added 13, 19, and 23 to the list. Another prime number is $2^{21701} - 1$ (verification of this is a difficult task!) This prime number is the largest *known* prime number as of January 1, 1979. You may wonder if there are larger prime numbers that have not yet been discovered. The answer is yes, for we will next show that there is no largest prime number. Therefore, there exist primes that are greater than $2^{21701} - 1$, although we do not know what they are. This is an old result, first proved by Euclid sometime around 300 B.C. Not only is it ancient but some people consider it one of the most elegant proofs in all of mathematics. We present it below essentially as Euclid proved it (except that Euclid wrote in Greek).

EXAMPLE. Before looking at Euclid's theorem, we might consider an example to see how you might go about showing that there is no largest prime number. We have stated that 17 is a prime number. Of course, it is not the largest prime number because by searching we quickly find that 19 is greater than 17 and 19 is a prime number. However, this method of examination shows us little; what we need is something that asserts the existence of prime numbers greater than 17 rather than simply exhibiting one. To emphasize what we are saying, you might try searching for the next largest prime after 6491. The key to our example is the construction of a number that is not divisible by any of the primes between 2 and 17. Such a number is

$$2 \cdot 3 \cdot 5 \cdot 7 \cdot 11 \cdot 13 \cdot 17 + 1 = 510{,}510 + 1 = 510{,}511$$

Since $1 = 510{,}511 - 510{,}510$, if any of 2, 3, 5, 7, 11, 13, or 17 were a divisor of 510,511, then because each of these also divides 510,510, we would obtain a positive divisor of 1 greater than 1, an impossibility. However, Theorem 3.1 asserts that 510,511 has prime divisors, and such a prime divisor is necessarily larger than 17.

With this example in mind, let us look at Euclid's theorem.

Theorem 3.2. There does not exist a largest prime number. (Stated differently, there are an infinite number of prime numbers.)

Proof. The plan is to assert that for each prime p there exists a prime number q which is larger than p. As in the example, let $p_1, p_2, \ldots, p_k = p$ denote all primes less than or equal to p, and let $N = p_1 p_2 \cdots p_k + 1$. By Theorem 3.1, N has a prime divisor, call it q. We claim that $q > p$. Indeed, if $q \leq p$, then $q = p_i$ for some $1 \leq i \leq k$. Now $q \mid N$ and $q \mid p_1 p_2 \cdots p_i \cdots p_k$, so by Lemma 2.1, $q \mid N - (p_1 p_2 \cdots p_k)$. Since $N - (p_1 p_2 \cdots p_k) = 1$ and $q > 1$, we have arrived at a contradiction, so $q > p$, as desired. ∎

In illustrating Theorem 3.1, we decomposed 420 in the following manner: $420 = 42 \cdot 10 = 7 \cdot 6 \cdot 2 \cdot 5 = 7 \cdot 2 \cdot 3 \cdot 2 \cdot 5$. We might have gone about decomposing 420 in a different fashion, such as $420 = 20 \cdot 21 = 4 \cdot 5 \cdot 21 = 2 \cdot 2 \cdot 5 \cdot 21 = 2 \cdot 2 \cdot 5 \cdot 3 \cdot 7$. Notice that in both cases, although their order varied, the primes involved were the same. That is, in both cases we ended up with two 2's, one 3, one 5, and one 7. We wish to prove that this is always the case. For convenience of discussion, let us agree to always arrange the primes involved in increasing order. For example, we will write $420 = 2 \cdot 2 \cdot 3 \cdot 5 \cdot 7$. To stress this idea, we make a formal definition.

Definition. For an integer $n > 1$, we say that a product of primes $p_1 p_2 \cdots p_k$ is a *prime factorization* of n if:

(i) $n = p_1 p_2 \cdots p_k$.

(ii) $p_1 \leq p_2 \leq \cdots \leq p_k$.

In terms of this definition, what we hope to prove is that each integer $n > 1$ has only one prime factorization. (Theorem 3.1 tells us that it has at least one prime factorization.) Thus if we are told that $p_1 p_2 \cdots p_t$ and $q_1 q_2 \cdots q_s$ are both prime factorizations of an integer n, we want to be able to conclude that $t = s$ and that $p_1 = q_1, p_2 = q_2, \ldots, p_t = q_s$. To do this we will need a simple but important fact about primes which we now prove.

Lemma 3.3. Let p be a prime number and let n be any integer. If $p \mid n$, then $(n, p) = p$. If $p \nmid n$, then $(n, p) = 1$.

Proof. If $p \mid n$, then p is a common divisor of n and p, and is clearly the largest such common divisor. Now suppose that $p \nmid n$ and let $d = (n, p)$. Then since $d \mid p$, we have that $d = 1$ or that $d = p$. However, $d \mid n$ and $p \nmid n$, so we cannot have $d = p$ and thus $d = 1$. ∎

Theorem 3.4. Let p be a prime number. If $p \mid ab$, then either $p \mid a$ or $p \mid b$.

Proof. If $p \mid a$, we are done. If $p \nmid a$, then by Lemma 3.3, $(a, p) = 1$, so by Lemma 2.12, $p \mid b$. ∎

Using the Principle of Induction, you should be able to prove the following consequence of Theorem 3.4.

Corollary 3.5. Let p be a prime number and $a_1, a_2, \ldots, a_n$ be integers. If $p \mid a_1 a_2 \cdots a_n$, then $p \mid a_i$ for some $1 \leq i \leq n$.

Theorem 3.6. If n is any integer greater than 1, then there is exactly one prime factorization for n.

Proof. Theorem 3.1 asserts that for each integer $n > 1$, there is at least one prime factorization and we wish to dispel the possibility that there be a different factorization. Since we wish to prove something about all integers $n \geq 2$, we turn to the Principle of Well-Ordering. Thus let $D = \{n \in \mathbb{Z} \mid n \geq 2$ and n has more than one prime factorization$\}$. As in the previous applications of the Principle of Well-Ordering, we would like to show that $D = \varnothing$. Since $n \geq 2$ for all $n \in D$, D does not have arbitrarily small integers, ruling out possibility (c) of the Principle of Well-Ordering. What about possibility (b)? Suppose that D has a least element and let us call it d. Then $d \geq 2$ and d has at least two distinct prime factorizations, say

$$d = p_1 p_2 \cdots p_t \qquad \text{and} \qquad d = q_1 q_2 \cdots q_s$$

where $p_1, \ldots, p_t, q_1, \ldots, q_s$ are prime numbers with $p_1 \leq p_2 \leq \cdots \leq p_t$ and $q_1 \leq q_2 \leq \cdots \leq q_s$. We claim that $p_1 = q_1$. Suppose that $p_1 \neq q_1$; then one of these two numbers is smaller than the other and we can assume that $p_1 < q_1$ (if $q_1 < p_1$, interchange the p's and q's in the following argument). Since $p_1 \mid d$ and $d = q_1 q_2 \cdots q_s$ we have that $p_1 \mid q_1 q_2 \cdots q_s$, so by Corollary 3.5, $p_1 \mid q_j$ for some $1 \leq j \leq s$. However, q_j, being a prime number, has no positive divisors other than 1 and q_j. Since $p_1 > 1$, we conclude that $p_1 = q_j$. However, $q_j \geq q_1$ by the manner in which the q's were arranged. Thus $p_1 = q_j \geq q_1 > p_1$, so $p_1 > p_1$. This contradiction shows that p_1 must equal q_1, as we claimed.

Once we have $p_1 = q_1$, from the equality $p_1 p_2 \cdots p_t = q_1 q_2 \cdots q_s$ we may cancel p_1 from both sides and obtain $p_2 p_3 \cdots p_t = q_2 q_3 \cdots q_s$. Let us call this number c. We have $1 \le c < d$. (Why is this true?) Since d was the least element in D, we must have $c \notin D$. Thus either $c < 2$ or else c has exactly one prime factorization. In the first case, since $c \ge 1$, we conclude that $c = 1$, which means that the prime factorizations for d had exactly one term: $d = p_1 = q_1$; but this would say that $d \notin D$, a contradiction. In the second case, we have at hand two prime factorizations for c, namely $c = p_2 p_3 \cdots p_t = q_2 q_3 \cdots q_s$. Since c has exactly one prime factorization we may conclude that $s = t$, $p_2 = q_2, \ldots, p_t = q_t$. Thus the two supposedly different prime factorizations for $d = p_1 p_2 \cdots p_t = q_1 q_2 \cdots q_t$ are really the same because we already have $p_1 = q_1$. This contradicts the requirement that for d these two prime factorizations were different. We have now eliminated possibility (b); hence $D = \varnothing$ and the proof is finished. ∎

Theorem 3.6 now enables us to speak about *the* prime factorization of an integer $n > 1$. For future purposes, it is convenient to group equal primes and use exponents to represent the factorization. For example, $24{,}750 = 2 \cdot 3 \cdot 3 \cdot 5 \cdot 5 \cdot 5 \cdot 11 = 2 \cdot 3^2 \cdot 5^3 \cdot 11$. Hence, for an integer $n > 1$, we write

$$n = p_1^{a_1} p_2^{a_2} \cdots p_t^{a_t}, \qquad \text{where } p_1 < p_2 < \cdots < p_t \text{ and } a_i \ge 1 \text{ for } 1 \le i \le t$$

We call this expression the *canonical representation of n* (from "canon" meaning "law" or "rule").

Knowing the prime factorization of an integer totally determines its divisors. For suppose that $n \ge 2$ and has canonical representation

$$n = p_1^{a_1} p_2^{a_2} \cdots p_k^{a_k}$$

If m is a positive divisor of n, then each prime divisor q of m is also a prime divisor of n, hence must be one of $p_1, \ldots, p_k$. In addition, the highest power b of q occurring as a factor of m must also occur as a factor of n; hence if $q = p_i$, then $b \le a_i$. Thus m must have the form

$$m = p_1^{b_1} p_2^{b_2} \cdots p_k^{b_k}$$

where $0 \le b_i \le a_i$. Notice that we allow the possibility of $b_i = 0$ so that $p_i^0 = 1$. For example, if $n = 3^2 \cdot 5^4 \cdot 13^7$ and $m = 15$, then $m \mid n$ and $m = 3 \cdot 5 = 3^1 \cdot 5^1 \cdot 13^0$. On the other hand, if $m = p_1^{b_1} p_2^{b_2} \cdots p_k^{b_k}$, where $0 \le b_i \le a_i$, then we have

$$\begin{aligned} n = p_1^{a_1} p_2^{a_2} \cdots p_k^{a_k} &= (p_1^{a_1 - b_1} \cdot p_1^{b_1})(p_2^{a_2 - b_2} \cdot p_2^{b_2}) \cdots (p_k^{a_k - b_k} \cdot p_k^{b_k}) \\ &= p_1^{a_1 - b_1} p_2^{a_2 - b_2} \cdots p_k^{a_k - b_k} \cdot m \end{aligned}$$

so $m \mid n$. Thus we have established the following:

Theorem 3.7. Let $n \geq 2$ be an integer with canonical representation $n = p_1^{a_1}p_2^{a_2}\cdots p_k^{a_k}$. An integer $m \geq 1$ is a positive divisor of n if and only if $m = p_1^{b_1}p_2^{b_2}\cdots p_k^{b_k}$, where $0 \leq b_i \leq a_i$ for all $1 \leq i \leq k$.

Corollary 3.8. If $n \geq 2$ is an integer with canonical representation $n = p_1^{a_1}p_2^{a_2}\cdots p_k^{a_k}$, then the number of positive divisors of n is given by the expression $(a_1 + 1)(a_2 + 1)\cdots(a_k + 1)$.

Proof. The proof of this corollary simply requires your counting up the possibilities for exponents, and we leave this as an exercise. (3.B.1) ∎

As another application of prime factorization we shall show you how greatest common divisors can be calculated once the prime factorizations are available.

EXAMPLE. If $n = 3^2 \cdot 5^3 \cdot 7 \cdot 13^2 \cdot 23$ and $m = 3 \cdot 7 \cdot 13^4 \cdot 17 \cdot 29$, then $(n, m) = 3 \cdot 7 \cdot 13^2$. To verify this, we check the two characterizing properties of the greatest common divisor given in Chapter 2. Let $d = 3 \cdot 7 \cdot 13^2$; then $d \mid n$ and $d \mid m$. Also, any divisor c of both n and m can have no primes different from 3, 7, or 13 if c is to divide both n and m. The fact that $c \mid m$ says that 3 can occur at most once as a factor of c; the fact that $c \mid n$ says that 7 occurs at most once as a factor of c and that 13 occurs at most twice. Thus $c \leq d$.

This process is quite straightforward to apply and we formalize it.

Theorem 3.9. Let n and m be positive integers with canonical representations $n = p_1^{a_1}\cdots p_k^{a_k}$, $m = q_1^{b_1}\cdots q_l^{b_l}$ and let $r_1, \ldots, r_t$ be the primes that occur in both representations (i.e., $\{r_1, \ldots, r_t\} = \{p_1, \ldots, p_k\} \cap \{q_1, q_2, \ldots, q_l\}$). If c_i is the least power of r_i that occurs in the representations of n and m, then

$$(n, m) = r_1^{c_1}r_2^{c_2}\cdots r_t^{c_t}$$

Proof. Let $d = r_1^{c_1}r_2^{c_2}\cdots r_t^{c_t}$. By Theorem 3.7, $d \mid n$ and $d \mid m$. So part (i) of the characterizing property of (n, m) is satisfied by d. For part (ii), let c be a common divisor of n and m. Any prime divisor p of c is a prime divisor of both n and m; hence $p \in \{r_1, \ldots, r_t\}$. Thus $c = r_1^{e_1}r_2^{e_2}\cdots r_t^{e_t}$, where $e_i \geq 0$. It remains to show that $e_i \leq c_i$ for $i = 1, \ldots, t$. This is so because if $1 \leq i \leq t$, then $r_i = p_v = q_u$ for some $1 \leq v \leq k$, $1 \leq u \leq l$; hence $c \mid n$ implies that $e_i \leq a_v$ while $c \mid m$ implies that $e_i \leq b_u$. Combining these with the definition of c_i, we conclude that $e_i \leq c_i$. Thus we have $c \leq d$, so part (ii) holds. Accordingly, $d = (n, m)$. ∎

3.2 A Sieve Method

So far we have used the existence of the prime factorizations but have not told you how to obtain them. Given an integer $n > 1$, if you wish to find the prime factorization of n, you first look for a prime divisor of n. The next theorem can be of some help in this search.

Theorem 3.10. If $n > 1$ is not a prime number, then n has a prime divisor $p \leq \sqrt{n}$.

Proof. n is not prime, so $n = ab$, where $1 < a < n$ and $1 < b < n$. If both $a > \sqrt{n}$ and $b > \sqrt{n}$, then $ab > (\sqrt{n})^2 = n$. Thus we must have $a \leq \sqrt{n}$ or $b \leq \sqrt{n}$; for definiteness, say $a \leq \sqrt{n}$. Then any prime divisor p of a satisfies $p \leq \sqrt{n}$ and is also a prime divisor of n. ∎

As an example, suppose that $n = 23{,}803$. Since $(154)^2 = 23{,}716$ and $(155)^2 = 24{,}025$, we have that

$$154 < \sqrt{23{,}803} < 155$$

Thus if 23,803 is not prime, it will have a prime divisor that is less than 155. Hence if you wished to find the prime factorization of 23,803, you would first test the primes less than 155 to see if 23,803 is divisible by any of these. Note that if none of these divides 23,803, then 23,803 is a prime number so that Theorem 3.10 can also serve in determining whether or not an integer is a prime.

Fortunately, tables exist that list the smallest prime divisor of an integer n, for $1 < n \leq 10{,}000$, as well as all primes $\leq 10^8$. It might be useful to have at hand a process for finding all primes less than a fixed positive integer K. One well-known process is quite ancient ($\sim$200 B.C.) and is known as the Sieve of Eratosthenes.

Let us begin by listing all integers from 2 to K. Circle 2, the first prime in the list. Then cross out all larger multiples of 2, namely 4, 6, 8, ... Then find the next number that is not crossed out, which is 3 (a prime) and circle it. Then cross out the larger multiples of 3, namely 6, 9, 12, ... Find the next number that is not crossed out, which is 5 (a prime), and circle it. Then cross out 10, 15, 20, ..., and continue this process.

We encourage you to try this, for example with $K = 170$. You will notice when you are done that the numbers you have circled are exactly the primes equal to or less than $K = 170$. We leave it as an exercise to show why this is true (2.B.19).

You may wonder if it is necessary to continue crossing out numbers until we reach K. Fortunately, Theorem 3.10 enables us to stop once we exceeded $\sqrt{K}$. This is because any nonprime $n \geq 2$ less than K has a prime divisor $\leq \sqrt{K}$, so if it were not prime, it would have been crossed out.

3.3 Fermat Factorization

We now wish to mention a method for finding all ways of factoring an odd integer into two unequal factors. The method was invented by Pierre de Fermat, a brilliant seventeenth-century mathematician whose work will be discussed repeatedly in this text. We will leave the justification of why the method works for the exercises. Here we only present the method.

STEP 1. If $n > 0$ is an odd integer, make a table consisting of $(n - 1)/2$ rows, obtained by successively adding consecutive odd integers to n. Note which rows of the table contain perfect squares.

STEP 2. If the sth row contains t^2, then $(t + s)(t - s) = n$ and all factorings of n into two distinct factors are produced in this manner.

ILLUSTRATION (for $n = 21$). In Table 3–1, the second row contains 5^2. Thus $(5 + 2)(5 - 2) = (7)(3) = 21$. Also, the tenth row contains 11^2. Thus $(11 + 10)(11 - 10) = (21)(1) = 21$. Note that these are the only such ways in which 21 can be factored [ignoring trivial variation such as $(3)(7) = 21$ or $(-7)(-3) = 21$].

Of course, it is silly to apply this method to a small number like 21. However, for a large n such as 100,117, it becomes a feasible method. All it requires is a bunch of additions and recognizing perfect squares. (Of course, having a table of squares at hand makes it very easy.)

Table 3-1 Illustration for $n = 21$

Row Number	
1	$21 + 1 = 22$
2	$22 + 3 = 25 = 5^2$
3	$25 + 5 = 30$
4	$30 + 7 = 37$
5	$37 + 9 = 46$
6	$46 + 11 = 57$
7	$57 + 13 = 70$
8	$70 + 15 = 85$
9	$85 + 17 = 102$
10	$102 + 19 = 121 = 11^2$

Note: $10 = (21 - 1)/2$.

On the other hand, you may argue that today such things are done by computers, so who cares if there is a 300-year-old "easy method"? Well, let us remind you that computer time is expensive, and programming a computer to use an easy method rather than some brute-force method is still very advantageous.

3.4 Some Unsolved Problems

In presenting prime numbers we have only given the beginnings of what is and has been a deep and fascinating aspect of mathematics. In the following chapters we will see how central prime numbers are to the understanding of number theory. Let us close this chapter by mentioning some of the unsolved problems that revolve around prime numbers. In doing so we will see what is perhaps one of the most intriguing aspects of number theory; that is, it is quite often easy to present problems which are so simple to state that any high school student can easily understand the question, and yet the solutions have eluded the sustained efforts of the finest mathematicians. We begin with Goldbach's Conjecture.

Goldbach's Conjecture. Every even integer greater than 2 can be written as the sum of two primes. (*Note:* Repetitions are permitted as in $6 = 3 + 3$.)

For quite a long period of time no progress at all was made on this conjecture. A few decades ago, however, it was shown that any even integer larger than $e^{e^{16.038}}$ could be written as the sum of at most four primes. (Recall that $e = 2.718\ldots$.) Nonetheless, the full conjecture remains unscathed. Modern computers have been of no help in solving this problem first posed over 230 years ago!

The Twin Primes Problem. There are some primes p such that $p + 2$ is also a prime. For instance, $p = 5$, $p + 2 = 7$; $p = 11$, $p + 2 = 13$; or $p = 29$, $p + 2 = 31$. These are examples of what are called *twin primes.* Are there an infinite number of such pairs? No one knows.

Mersenne Primes For some primes p, $2^p - 1$ is also a prime. This occurs with $p = 5, 7, 13, 19{,}937$. Prime numbers of the form $2^p - 1$, where p is prime are called *Mersenne primes.* Mersenne was a Jesuit priest of the seventeenth century who served as a one-man clearinghouse for mathematical advances, back in the days when such an activity was possible. It is not known whether or not there are an infinite number of Mersenne primes. The prime $2^{21701} - 1$ is the largest known Mersenne prime. These primes will turn up again in Chapter 7 in relation to yet another famous problem.

$n^2 + 1$. Are there infinitely many primes of this form?

Feel free to try your hand at any of these problems. But be warned: Whatever progress has been made on these problems, or whatever solution has been found to a problem of a similar nature, has quite often required extremely difficult and sophisticated techniques. At times these techniques had to be invented solely for the purpose of attacking the problem at hand. In short, it is safe to say that number theory, while posing problems that appear to be seductively simple, is in truth one of the most difficult undertakings of the human mind.

EXERCISES 3.A

1. Find prime decompositions for each of the following integers.
   (a) 49,500 (b) 12,500
   (c) 38,712 (d) 10,083
2. For a positive integer n, let $n!$ denote the product $1 \cdot 2 \cdot 3 \cdots (n-2)(n-1)n$ (for example, $4! = 1 \cdot 2 \cdot 3 \cdot 4 = 24$). Find the prime decomposition of $7! + 1$ and verify that each prime divisor is greater than 7.
3. Give an example of integers m and n such that $m \nmid n$ but $(m, n) \neq 1$. Does this violate the second sentence in Lemma 3.3?
4. Show by example that the condition that p is a prime is necessary in Theorem 3.4.
5. Let $a > 1$ be an integer such that $12 \mid a$. Show that 2 and 3 occur in the prime factorization of a.
6. Given an integer $a > 1$ how are the prime factorizations of a and a^2 related? (*Hint:* Try some examples.)
7. Find all the positive divisors for each of the numbers in Exercise 3.A.1.
8. In Exercise 3.A.7, verify that the number of positive divisors for each integer is in agreement with the number predicted by Corollary 3.8.
9. Find the smallest positive integer with six positive divisors; eight positive divisors.
10. Let p and q be distinct primes such that $p \mid a$ and $q \mid a$. Show that $pq \mid a$. (*Hint:* Look at the prime decomposition of a).
11. Use prime decompositions to calculate (49500, 12500) and (38712, 12500).
12. Use prime decompositions to calculate [49500, 12500] and [38712, 12500]. (See Exercise 2.B.6 for the notation $[a, b]$.)
13. Prove Exercise 2.B.6(b) using prime decompositions.
14. Apply the Sieve of Eratosthenes to find all primes less than 170.
15. How can the Sieve of Eratosthenes be used to find all prime divisors of 168?

16. Use Fermat's factorization technique to find all ways of factoring $n = 63$ into two factors. Do the same for $n = 126$.

17. Describe how you could use Fermat's factorization technique to find the prime decomposition of a given integer.

18. Verify that every even integer between 4 and 100 can be expressed as the sum of two primes.

19. Find all pairs of twin primes less than 100.

20. Verify that $2^5 - 1$ and $2^7 - 1$ are primes, while $2^{11} - 1$ is not.

21. Determine whether or not $n \mid (n - 1)! + 1$ for $n = 3, 4, 5, 6, 7, 8, 9, 10, 11,$ 12, and 13. Can you guess a general result?

22. In Chapter 2, we mentioned that $3 \mid a^3 - a$ for any integer a. Convince yourself by looking at examples that $5 \mid a^5 - a$ and $7 \mid a^7 - a$ appear to be true statements. (*Note:* They *are* true.) What about the statement $6 \mid a^6 - a$?

23. Use prime decompositions to show that if $(a, c) = 1$ and $(b, c) = 1$ then $(ab, c) = 1$. (Also see Exercise 2.B.9.)

24. Prove Theorem 3.1 using the Principle of Well-Ordering.

25. Prove Theorem 3.6 using the Principle of Induction.

EXERCISES 3.B

1. Prove Corollary 3.8.

2. Show that any prime of the form $3k + 1$ is also of the form $6n + 1$.

3. (a) Show that none of the four consecutive numbers

$$5! + 2, \qquad 5! + 3, \qquad 5! + 4, \qquad 5! + 5$$

is a prime number.

(b) Show that none of the six consecutive numbers $7! + (k + 1)$, $1 \leq k \leq 6$, is a prime number.

(c) Deduce a general result based on (a) and (b).

4. Prove that every integer of the form $4k + 3$ has a prime divisor of the form $4k + 3$. (*Hint:* Observe that any prime greater than 2 is of the form $4n + 1$ or $4n + 3$.)

5. Use Exercise 3.B.4 to show that there are an infinite number of primes of the form $4k + 3$.

6. Determine all integers n such that n, $n + 2$, and $n + 4$ are all primes.

7. Let a and b be nonnegative integers not both 1. Show that if k is an odd positive integer, $k \geq 3$, then $a^k + b^k$ is not prime.

8. Find the smallest positive value of n such that $n^2 + n + 41$ is not prime. (*Hint:* The table of primes may be useful.)

9. Show that if $n^2 + 2$ is prime, then $3 \mid n$.

10. Prove: If $2^k + 1$ is a prime for some integer $k \geq 1$, then $k = 2^t$ for some $t \geq 1$. (*Hint:* Write $k = 2^t m$, where $2 \nmid m$. Apply Exercise 3.B.7.)

11. Show that if $a^n - 1$ is prime, then n is prime. (*Hint:* If $n = lk$, use long division to divide $a^n - 1$ by $a^l - 1$.)

12. Show that the only prime p such that $3p + 1$ is a perfect square is $p = 5$.

13. (a) Show that the equation $3n + 2 = m^2$ has no integer solutions.
(b) Show that the equation $5n + 3 = m^2$ has no integer solutions.

14. Suppose that $a, b, c \in Z$ with $a > 0, b > 0$:
(a) Show that $[a, b] \mid c$ if and only if $a \mid c$ and $b \mid c$.
(b) Show that if either $c \mid a$ or $c \mid b$, then $c \mid [a, b]$.
(c) Give an example where $c \mid [a, b]$ but $c \nmid a$ and $c \nmid b$.
(d) Show that if $c = p^e$ with p prime and $e > 0$, then $c \mid [a, b]$ if and only if either $c \mid a$ or $c \mid b$.
(For the notation $[a, b]$, see Exercise 2.B.6.)

The following three problems are concerned with showing why Fermat's factorization method works.

15. Show in the table you construct while using Fermat's factorization method that the entry in the kth row is $n + k^2$. [*Hint:* Exercise 1.A.4(h).]

16. Show that if $a > b > 0$ are integers and if $ab = n$, then

$$\left(\frac{a+b}{2}\right)^2 - \left(\frac{a-b}{2}\right)^2 = n$$

17. Use the preceding two exercises to explain why Fermat's factorization technique does give all ways of factoring a nonsquare odd integer $n > 0$ into two distinct factors.

18. Let $n > 1$ be a nonsquare odd integer. Show that n is prime if and only if the smallest positive value of k for which $n + 1 + 3 + 5 + 7 + \cdots + (2k - 1)$ is a perfect square is $k = (n - 1)/2$.

19. Prove that the Sieve of Eratosthenes finds all primes (and only primes) between 2 and the given integer k.

4

Congruences

ONE of the beauties of mathematics is that many times a seemingly simple idea can have far-reaching effects, and it is the study of one such idea to which we now turn. Let us begin by considering the following problem:

Show that the equation

$$x^7 + x^5 = 3{,}226{,}583{,}002{,}173{,}942{,}667{,}382{,}111{,}913{,}613{,}525{,}837$$

has no integer solutions.

A trial-and-error attack would be quite lengthy (and expensive) even for a large computer. Instead, we might reason as follows. Every integer is either even or odd. If x is an even integer, then x^7 is even, as is x^5. Since the sum of two even integers is again even, $x^7 + x^5$ will be even. As the number on the right-hand side is odd, no even integer can be a solution of the equation. What if x is any odd integer? In this case x^7 and x^5 are both odd integers but the sum of two odd integers is even, so again $x^7 + x^5$ is even, showing that no odd integer x can be a solution either. Therefore, there can be no integer solutions to the equation. The idea embodied in this example generalizes to a very useful concept.

4.1 Congruence

In this argument we have used the fact that every integer is either even or odd. Thus the set $\mathbb{Z}$ can be split up into two subsets which are reasonably well behaved. We can describe these two sets in a manner that may suggest a possible generalization. The set of even integers is the set

$$E = \{2k \mid k \in \mathbb{Z}\}$$

and the set of odd integers is the set

$$0 = \{2k + 1 \mid k \in \mathbb{Z}\}$$

Suppose that we do the same with a positive integer other than 2, say 6. We then have

$$\{6k \mid k \in \mathbb{Z}\} = \{\ldots, -12, -6, 0, 6, 12, \ldots\}$$

and

$$\{6k + 1 \mid k \in \mathbb{Z}\} = \{\ldots, -11, -5, 1, 7, 13, \ldots\}$$

In this case (unlike the case with 2) these two sets together do not exhaust all of $\mathbb{Z}$. For example, neither 2 nor 3 is in either set. More to the point, we also have the sets

$$\begin{aligned}\{6k + 2 \mid k \in \mathbb{Z}\} &= \{\ldots, -10, -4, 2, 8, 14, \ldots\}\\ \{6k + 3 \mid k \in \mathbb{Z}\} &= \{\ldots, -9, -3, 3, 9, 15, \ldots\}\\ \{6k + 4 \mid k \in \mathbb{Z}\} &= \{\ldots, -8, -2, 4, 10, 16, \ldots\}\\ \{6k + 5 \mid k \in \mathbb{Z}\} &= \{\ldots, -7, -1, 5, 11, 17, \ldots\}\end{aligned}$$

You can easily convince yourself that these six sets do indeed exhaust $\mathbb{Z}$. Notice also that no two of them have any integers in common.

Now suppose that we consider an arbitrary positive integer m. Relying on the two previous examples, we will have m sets:

$$(*)\qquad \begin{aligned}\{km \mid k \in \mathbb{Z}\} &= \{\ldots, -2m, -m, 0, m, 2m, \ldots\}\\ \{km + 1 \mid k \in \mathbb{Z}\} &= \{\ldots, -2m + 1, -m + 1, 1, m + 1, 2m + 1, \ldots\}\\ \{km + 2 \mid k \in \mathbb{Z}\} &= \{\ldots, -2m + 2, -m + 2, 2, m + 2, 2m + 2, \ldots\}\\ &\vdots\\ \{km + (m - 1) \mid k \in \mathbb{Z}\} &= \{\ldots, -2m + (m - 1), -m + (m - 1), m - 1,\\ &\qquad m + (m - 1), 2m + (m - 1), \ldots\}\end{aligned}$$

The Division Algorithm (Theorem 2.4) asserts that every integer is in one of these sets; hence these m sets exhaust $\mathbb{Z}$. The uniqueness of the quotient and remainder simply means that no two distinct sets have anything in common. Let's examine each of these sets in $(*)$ a bit closer.

Suppose that $0 \leq r \leq m - 1$ and a and b are both members of the set $\{km + r \mid k \in \mathbb{Z}\}$. Then $a = k_1 m + r$ and $b = k_2 m + r$ for some integers k_1 and k_2. Hence

$$a - b = (k_1 m + r) - (k_2 m + r) = (k_1 - k_2)m, \qquad \text{showing that } m \mid a - b$$

Thus if a and b are in the same set, $m \mid a - b$. The converse also holds. For suppose that a and b are integers such that $m \mid a - b$. We first have $b = km + r$ for some $k \in \mathbb{Z}$, $0 \leq r \leq m - 1$. Since $m \mid a - b$, there exists $n \in \mathbb{Z}$ such that $a - b = mn$. Thus $a = b + mn = (km + r) + (mn) = (k + n)m + r$, showing that a and b are in the same set in $(*)$. To summarize: Two integers fall in the same set in $(*)$ if and only if their difference is divisible by m.

This last comment inspires our next definition. First though, notice that $\{mk \mid k \in Z\}$ is the ideal of all multiples of m. We only take $m > 0$ since $\{(-m)k \mid k \in Z\} = \{mk \mid k \in Z\}$ and $\{0k \mid k \in Z\} = \{0\}$ is dull.

Definition. Let m be a positive integer. If $a, b \in \mathbb{Z}$, we say that *a is congruent to b modulo m* provided that $m \mid (a - b)$.

Notation. We symbolize the statement "a is congruent to b modulo m" by writing: $a \equiv b \bmod m$. If a is *not* congruent to b modulo m, we write $a \not\equiv b \bmod m$.

Using $m = 6$, we have $8 \equiv 2 \bmod 6$ because $6 \mid 8 - 2$. As another example, $3715 \equiv -3281 \bmod 6$ because $3715 - (-3281) = 6996$ and $6 \mid 6996$. Since $6 \nmid 7 - 2$, we have $7 \not\equiv 2 \bmod 6$.

It is no coincidence that the symbol used to denote congruence resembles the symbol used to denote equality, for congruence has most (but not all) of the properties that equality has. Our first theorem of this chapter will show you that congruence behaves very much like equality and provides you with some guidelines for working with congruence.

Theorem 4.1. Let m be a positive integer.

(i) For any integer a, $a \equiv a \bmod m$.
(ii) If $a \equiv b \bmod m$ and $c \equiv d \bmod m$, then $a + c \equiv b + d \bmod m$.
(iii) If $a \equiv b \bmod m$ and $c \equiv d \bmod m$, then $ac \equiv bd \bmod m$.
(iv) If $c \equiv d \bmod m$, then $-c \equiv -d \bmod m$.
(v) If $a \equiv b \bmod m$ and $c \equiv d \bmod m$, then $a - c \equiv b - d \bmod m$.
(vi) If $a \equiv b \bmod m$ and $b \equiv c \bmod m$, then $a \equiv c \bmod m$.
(vii) If $a \equiv b \bmod m$, then $b \equiv a \bmod m$.

Proof. We shall use the definition of congruence to prove (i), (ii), (iii), (vi), and (vii).

(i) For any integer a, $a - a = 0$ and $m \mid 0$. Thus $a \equiv a \bmod m$.
(ii) $a \equiv b \bmod m$ means that $m \mid a - b$; $c \equiv d \bmod m$ means that $m \mid c - d$. By Lemma 2.1, $m \mid (a - b) + (c - d)$. However $(a - b) + (c - d) = (a + c) - (b + d)$, so $m \mid (a + c) - (b + d)$, showing that $a + c \equiv b + d \bmod m$.
(iii) Again $m \mid (a - b)$ and $m \mid (c - d)$; we must show that $m \mid ac - bd$. However, $ac - bd = ac - bc + bc - bd = (a - b)c + b(c - d)$. By Corollary 2.3, $m \mid ac - bd$.

(iv) This is a consequence of (i) and (iii), since $-1 \equiv -1 \bmod m$ and $-c = -1 \cdot c$, $-d = -1 \cdot d$.

(v) This one follows from (ii) and (iv).

(vi) We have $m \mid a - b$ and $m \mid b - c$; hence $m \mid (a - b) + (b - c)$. Thus $m \mid a - c$, so $a \equiv c \bmod m$.

(vii) Do this as an exercise. ∎

EXAMPLE. Next we give an example of how the various parts of the theorem can be used. Again let $m = 6$ and consider the expression $3 \cdot 19 + 8^3 - 11 \cdot 14 = N$. Let us find a number r between 0 and 5 such that $N \equiv r \bmod 6$.

(a) First, $19 \equiv 1 \bmod 6$ and by (i), $3 \equiv 3 \bmod 6$, so by (iii) $3 \cdot 19 \equiv 3 \cdot 1 \bmod 6$, hence $3 \cdot 19 \equiv 3 \bmod 6$.

(b) Next, $8 \equiv 2 \bmod 6$, so using (iii) with $a = c = 8$ and $b = d = 2$, we have $8^2 \equiv 4 \bmod 6$. Using (iii) once more yields

$$8^2 \equiv 4 \bmod 6 \qquad \text{and} \qquad 8 \equiv 2 \bmod 6$$

Hence $8^3 \equiv 8 \bmod 6$. Now, using (vi) we obtain $8^3 \equiv 8 \bmod 6$ and $8 \equiv 2 \bmod 6$, so $8^3 \equiv 2 \bmod 6$.

(c) We have $11 \equiv 5 \bmod 6$ and $14 \equiv 2 \bmod 6$, so $11 \cdot 14 \equiv 10 \bmod 6$ by (iii). Since $10 \equiv 4 \bmod 6$, we have $11 \cdot 14 \equiv 4 \bmod 6$.

(d) The final step is to combine (a), (b), and (c) using (ii) and (v). Thus

$$3 \cdot 19 \equiv 3 \bmod 6, \qquad 8^3 \equiv 2 \bmod 6, \qquad 11 \cdot 14 \equiv 4 \bmod 6$$

so $(3 \cdot 19 + 8^3) - 11 \cdot 14 \equiv (3 + 2) - 4 \bmod 6$.
Hence

$$3 \cdot 19 + 8^3 - 11 \cdot 14 \equiv 1 \bmod 6.$$

A comparison of Theorem 4.1 with the properties listed at the beginning of Chapter 1 shows the similarity of congruence and equality. There is one significant way in which congruence differs from equality, and this difference concerns rule 6°, which states that if $a, b \in \mathbb{Z}$ and $ab = 0$, either $a = 0$ or $b = 0$. An equivalent way of stating 6° is as cancellation: Given integers a, b, and c, if $ab = ac$ and $a \neq 0$, then $b = c$. For congruences the parallel statement is just not true; that is, it is possible to have $ab \equiv ac \bmod m$ and $a \not\equiv 0 \bmod m$ but $b \not\equiv c \bmod m$. For example, take $m = 12$, $a = 8$, $b = 2$, and $c = 5$. Certainly, $8 \cdot 2 \equiv 8 \cdot 5 \bmod 12$ and $8 \not\equiv 0 \bmod 12$ but $2 \not\equiv 5 \bmod 12$. Thus cancellation of a common factor in congruences is not always possible. However under certain circumstances we may still safely "cancel."

Theorem 4.2. Let $m > 0$ and suppose that $ab \equiv ac \bmod m$. If $(a, m) = 1$, then $b \equiv c \bmod m$.

Proof. We have $ab \equiv ac \bmod m$, so that $m \mid ab - ac$. As $ab - ac = a(b - c)$, we have $m \mid a(b - c)$. Now $(a, m) = 1$, so using Lemma 2.12 we have $m \mid b - c$, hence $b \equiv c \bmod m$. ∎

In order to help you see the relationship between equality and congruence, and also as a preview of some of the things that we will encounter, let us consider the equation $5X + 26Y = 9$. Rewrite this as $26Y = 9 - 5X$. If we have a solution x, y to this equation, $26y = 9 - 5x$ says that $26 \mid 9 - 5x$. Conversely, if $26 \mid 9 - 5x$ for some x, then there is an integer y such that $26y = 9 - 5x$ and so the pair x, y is a solution of the equation $5X + 26Y = 9$. Thus the problem of finding integers x and y such that $5x + 26y = 9$ is the same as finding an integer x such that $26 \mid 9 - 5x$ and then letting $y = (9 - 5x)/26$. Now, however, an equivalent formulation of $26 \mid 9 - 5x$ is $5x \equiv 9 \bmod 26$. This means that the problem of solving the Diophantine equation $5X + 26Y = 9$ is in essence the same as solving the congruence $5X \equiv 9 \bmod 26$ (or the congruence $26Y \equiv 9 \bmod 5$ if we interchange the roles of x and y). In Chapter 5 we thoroughly investigate the congruence $ax \equiv c \bmod m$ using this idea.

4.2 Inverses Modulo *m*

For the present we will consider one aspect of this problem which will partially motivate some of our work in the present chapter. Let us look at the linear equation $5x = 9$. If you were asked to solve this equation you would proceed to make the coefficient of x equal to 1 by multiplying the equation by $\frac{1}{5}$, getting $x = \frac{9}{5}$. Since $\frac{9}{5} \notin \mathbb{Z}$, you would say that no integer solutions exist. For the congruence $5x \equiv 9 \bmod 26$, we can perform a similar operation. Notice that $5 \cdot 21 = 105$ and $105 \equiv 1 \bmod 26$. Therefore, if we multiply $5x \equiv 9 \bmod 26$ by 21, we get $5 \cdot 21x \equiv 9 \cdot 21 \bmod 26$. However, $5 \cdot 21 \equiv 1 \bmod 26$ and $9 \cdot 21 = 189 \equiv 7 \bmod 26$, so this last congruence becomes $1 \cdot x \equiv 7 \bmod 26$. Thus the integers x which satisfy the congruence $5x \equiv 9 \bmod 26$ are the members of the set $\{x \in \mathbb{Z} \mid x \equiv 7 \bmod 26\} = \{x \in \mathbb{Z} \mid x = 26k + 7, \text{ where } k = 0, \pm 1, \pm 2, \ldots\} = \{\ldots, -45, -19, 7, 33, \ldots\}$.

The previous example was solvable because we were able to find a number c (in the example, $c = 21$) such that $5c \equiv 1 \bmod 26$. This may not always be possible. For instance, there is no integer c such that $6c \equiv 1 \bmod 8$, as we will soon prove, so this technique is not immediately helpful in trying to solve a congruence such as $6x \equiv 4 \bmod 8$. In Chapter 5 we will see how to deal with this problem.

It is then clearly of interest to be able to answer the following question: Given integers a, m with $m > 0$, when is there an integer c such that $ac \equiv 1 \bmod m$?

Theorem 4.3. Let $a, m \in \mathbb{Z}$ with $m > 0$. Then there is an integer c such that $ac \equiv 1 \bmod m$ if and only if $(a, m) = 1$.

Proof. Suppose first that there exists $c \in \mathbb{Z}$ such that $ac \equiv 1 \bmod m$. Then $m \mid ac - 1$; hence there is an integer k such that $ac - 1 = mk$. Thus $ac - mk = 1$. If $d = (a, m)$, then $d \mid a$ and $d \mid m$, so $d \mid 1$ by Corollary 2.3. Since $d \geq 1$, we must have $d = 1$.

Conversely, suppose that $(a, m) = 1$. Then by Theorem 2.8 there are integers u and v such that $au + mv = 1$. Therefore, $au - 1 = mv$, so $m \mid au - 1$, giving $au \equiv 1 \bmod m$. We let $c = u$ and we are done. ∎

EXAMPLE. Looking again at $a = 5, m = 26$, we had $c = 21$ since $5 \cdot 21 \equiv 1 \bmod 26$. However, 21 is not the only possibility. For example, $5 \cdot (-5) \equiv 1 \bmod 26$ and $5 \cdot 73 \equiv 1 \bmod 26$. Notice, though, that $-5 \equiv 21 \bmod 26$ and $73 \equiv 21 \bmod 26$ and so, modulo 26, 21 is not a different answer than -5 or 73.

Theorem 4.4. Let $m > 0$ and $a \in \mathbb{Z}$ with $(a, m) = 1$. Then there is a unique c modulo m such that $ac \equiv 1 \bmod m$. (By "unique modulo m" we mean that if $ac \equiv 1 \bmod m$ and $ac_1 \equiv 1 \bmod m$, then $c \equiv c_1 \bmod m$.)

Proof. From Theorem 4.3 we know that there exists at least one integer c such that $ac \equiv 1 \bmod m$. Suppose that c_1 is also an integer for which $ac_1 \equiv 1 \bmod m$. Then $ac \equiv 1 \equiv ac_1 \bmod m$. Since $(a, m) = 1$, Theorem 4.2 permits us to cancel a, leaving $c \equiv c_1 \bmod m$. ∎

What this theorem tells us is that modulo 26, 21 is the only integer such that $5 \cdot 21 \equiv 1 \bmod 26$. It therefore makes sense to refer to 21 as *the multiplicative inverse of* 5 *modulo* 26. (We may also refer to 5 as the multiplicative inverse of 21 mod 26.) One thing you should keep in mind regarding this terminology; 5 has a multiplicative inverse modulo 26 but not modulo 10, for example. Thus in speaking about multiplicative inverses, it is important to keep in mind the modulus m.

Given an integer $m > 0$ and $a \in \mathbb{Z}$ such that $(a, m) = 1$, the proof of Theorem 4.3 provides us with a means for finding the multiplicative inverse of a modulo m. We use the Euclidean algorithm to find integers u and v for which $au + mv = 1$, and then take $c = u$, or for that matter any integer congruent to u modulo m. What follows now is another way of determining c. The method embodies many ideas.

EXAMPLE. We return to our example of $a = 5$ and $m = 26$, calculating the powers of 5 modulo 26. Thus we have

$$5^0 = 1 \equiv 1 \bmod 26, \qquad 5^1 \equiv 5 \bmod 26, \qquad 5^2 \equiv 25 \bmod 26,$$
$$5^3 \equiv 125 \equiv 21 \bmod 26$$
$$5^4 \equiv 105 \equiv 1 \bmod 26, \qquad 5^5 \equiv 5 \bmod 26, \qquad 5^6 \equiv 25 \bmod 26,$$
$$5^7 \equiv 21 \bmod 26$$
$$5^8 \equiv 1 \bmod 26, \qquad 5^9 \equiv 5 \bmod 26, \ldots$$

A cyclic pattern emerges, namely, the integers 1, 5, 25, and 21 are repeated in that order. Thus, modulo 26, there are only four distinct powers of 5. Note also that 4 is the first positive integer for which a power of 5 is congruent to 1 modulo 26 and that $5^3 \equiv 21 \bmod 26$ is the multiplicative inverse of 5 modulo 26. Keeping these things in mind, it is easy to see that for any integer $k \geq 0$ we have

$$5^{4k} \equiv 1 \bmod 26, \qquad 5^{4k+1} \equiv 5 \bmod 26, \qquad 5^{4k+2} \equiv 25 \bmod 26,$$
$$5^{4k+3} \equiv 21 \bmod 26$$

With this example in mind, we first prove

Theorem 4.5. Let $a, m \in \mathbb{Z}$ with $m > 0$. Then there exists a positive integer d such that $a^d \equiv 1 \bmod m$ if and only if $(a, m) = 1$.

Proof. If there is an integer d such that $a^d \equiv 1 \bmod m$, then $c = a^{d-1}$ serves as a multiplicative inverse for a modulo m. Consequently, by Theorem 4.3, $(a, m) = 1$.

Now assume that $(a, m) = 1$. Each of the integers $a, a^2, \ldots, a^m$ is congruent modulo m to a number between 1 and $m - 1$. Now there are m different powers of a and only $m - 1$ distinct numbers between 1 and $m - 1$. This means that there must be at least two distinct powers of a which are congruent to the same number modulo m between 1 and $m - 1$. Let us say that a^j and a^l are congruent to the same number modulo m, where $j \neq l$. Then $a^j \equiv a^l \bmod m$. Either $j > l$ or $l > j$, so for definiteness suppose that $j > l$. Then $j - l > 0$ and $a^j = a^{j-l+l} = a^{j-l} \cdot a^l$; hence

$$a^{j-l} \cdot a^l \equiv a^l \bmod m$$

As $(a, m) = 1$ it follows that $(a^l, m) = 1$. By Theorem 4.2 we are permitted to cancel a^l and arrive at

$$a^{j-l} \equiv 1 \bmod m$$

Hence $d = j - l > 0$ works. ∎

In view of Theorem 4.5, we are led to the following definition.

Definition. Let $m > 0$ and suppose that $(a, m) = 1$. The least positive integer d such that $a^d \equiv 1 \bmod m$ is called *the order of a modulo m.*

Notice that Theorem 4.5 ensures that there is an integer $k > 0$ such that $a^k \equiv 1 \bmod m$ whenever $(a, m) = 1$. The Principle of Well-Ordering ensures that there is a smallest positive integer with this property.

A basic property of the order of a modulo m is reflected in the next proposition.

Theorem 4.6. Let $m > 0$, $a \in \mathbb{Z}$ with $(a, m) = 1$, and let d be the order of a modulo m. For any integer $k \geq 1$, $a^k \equiv 1 \bmod m$ if and only if $d \mid k$.

Proof. If $k = dt$, then

$$a^k \equiv a^{dt} \equiv (a^d)^t \equiv 1^t \equiv 1 \bmod m$$

On the other hand, suppose that $a^k \equiv 1 \bmod m$. Write $k = dt + r$, where $0 \leq r < d$. Then

$$1 \equiv a^k \equiv a^{dt+r} \equiv a^{dt} \cdot a^r \equiv a^r \bmod m$$

since $a^{dt} \equiv 1 \bmod m$. Thus r is a nonnegative integer less than d with the property that $a^r \equiv 1 \bmod m$. By the definition of the order of a modulo m, we are forced to conclude that $r = 0$. Hence $k = dt$, so $d \mid k$. ∎

The cyclic nature reflected by the powers of 5 modulo 26 holds in general.

Theorem 4.7. Let $m > 0$, $a \in \mathbb{Z}$ with $(a, m) = 1$, and let d be the order of a modulo m. Then $a^0 = 1, a^2, a^3, \ldots, a^{d-1}$ represent the distinct powers of a modulo m, and for any integer $k \geq 1$, $a^{kd+i} \equiv a^i \bmod m$ for any $0 \leq i \leq d - 1$.

Proof. We can argue as follows to show that the powers are distinct. Suppose that $0 \leq i \leq d - 1$ and $0 \leq j \leq d - 1$ and $a^i \equiv a^j \bmod m$. Either $i \geq j$ or $j \leq i$, so let us assume we have $i \geq j$. Then

$$a^j \equiv a^i \equiv a^{i-j} \cdot a^j \bmod m$$

Since $(a, m) = 1$, we have $(a^j, m) = 1$, so that $a^{i-j} \equiv 1 \bmod m$. As $0 \leq i - j < d$, the definition of order tells us that $i - j = 0$, hence $i = j$. Thus $a^0 = 1$, $a^2 \cdots a^{d-1}$ are distinct modulo m. Finally, $a^{kd+i} \equiv a^i \bmod m$ since $a^d \equiv 1 \bmod m$ implies that $a^{kd} \equiv 1 \bmod m$. ∎

4.3 Fermat's Little Theorem

We now have presented two ways of finding the multiplicative inverse c of an integer a modulo m where $(a, m) = 1$. The first method requires using the Euclidean algorithm to find integers u and v such that $au + mv = 1$ and then taking $c = u$. The second method requires finding the order d of a modulo m, then taking $c = a^{d-1}$. It would be nice to have some way of selecting candidates for d other than trying the integers $1, 2, \ldots, m$ one at a time. Fortunately, there is an elegant result available, known as Euler's Theorem, which among many other things serves as an effective screening agent. It will tell us where to look for the order. A special case of Euler's Theorem is known as *Fermat's Little Theorem*. We will present it first.

Definition. Let $m > 0$ and let $\{r_1, \ldots, r_k\}$ and $\{s_1, \ldots, s_k\}$ be two sets of integers each containing the same (finite) number of distinct elements. We say that $\{r_1, \ldots, r_k\}$ is congruent to $\{s_1, \ldots, s_k\}$ modulo m, written $\{r_1, \ldots, r_k\} \equiv \{s_1, \ldots, s_k\} \bmod m$, if there is a rearrangement of $r_1, \ldots, r_k$, call it $t_1, \ldots, t_k$, such that $t_1 \equiv s_1 \bmod m$, $t_2 \equiv s_2 \bmod m, \ldots, t_k \equiv s_k \bmod m$.

EXAMPLE. With $m = 6$ we have $\{3, 1\} \equiv \{7, 27\} \bmod 6$, since rearranging 3, 1 as 1, 3, we get $1 \equiv 7 \bmod 6$ and $3 \equiv 27 \bmod 6$. Similarly, $\{1, 2, 11, -2\} \equiv \{5, 8, -14, 1\} \bmod 6$, since rearranging $1, 2, 11, -2$ as $11, 2, -2, 1$ we get $11 \equiv 5 \bmod 6$, $2 \equiv 8 \bmod 6$, $-2 \equiv -14 \bmod 6$, and $1 \equiv 1 \bmod 6$. On the other hand, $\{1, 2, 3\} \not\equiv \{8, 7\} \bmod 6$, since the sets contain different numbers of elements. Also, $\{1, 2, 3\} \neq \{8, 7, 10\} \bmod 6$, since 3 is not congruent to any of 8, 7, or 10 mod 6. Finally, $\{2, 8\} \not\equiv \{14, 15\} \bmod 6$, since 15 is not congruent to either 2 or 8 mod 6.

Lemma 4.8. If $m > 0$ and if $\{r_1, r_2, \ldots, r_k\} \equiv \{s_1, s_2, \ldots, s_k\} \bmod m$, then $r_1 r_2 \cdots r_k \equiv s_1 s_2 \cdots s_k \bmod m$.

Proof. We are told that there is some rearrangement of $r_1, r_2, \ldots, r_k$, call it $t_1, t_2, \ldots, t_k$ such that $t_j \equiv s_j \bmod m$ for $j = 1, \ldots, k$. Now $r_1 r_2 \cdots r_k = t_1 t_2 \cdots t_k$ since multiplication is commutative. Finally, $t_1 t_2 \cdots t_k \equiv s_1 s_2 \cdots s_k$ by Theorem 4.1. ∎

EXAMPLE. $\{3, 1\} \equiv \{7, 27\} \bmod 6$. Now $3 \cdot 1 = 1 \cdot 3$ and since $1 \equiv 7 \bmod 6$ and $3 \equiv 27 \bmod 6$, we have $3 \cdot 1 = 1 \cdot 3 \equiv 7 \cdot 27 \bmod 6$.

Our next lemma is a useful technicality.

Lemma 4.9. Let i and j be integers with $1 \leq i \leq m$ and $1 \leq j \leq m$. If $i \equiv j \bmod m$, then $i = j$.

Proof. Suppose that $i \neq j$. Then we may assume that $i > j$. Obviously, $1 \leq i - j < m$. However, Exercise 1.A.1(e) shows that a positive integer cannot divide a smaller positive integer. This contradicts that $m \mid i - j$. Thus $i = j$. ∎

Lemma 4.10. Let $m > 0$ and suppose that $a \in Z$ with $(a, m) = 1$. Then $\{1 \cdot a, 2 \cdot a, \ldots, (m - 1)a\} \equiv \{1, 2, \ldots, m - 1\} \bmod m$.

Proof. Our goal is to show that $1 \cdot a, 2 \cdot a, \ldots, (m - 1)a$ can be rearranged so as to satisfy the definition of set congruence. By the division algorithm, for $i = 1, 2, \ldots, (m - 1)$ we may write $i \cdot a = q_i m + r_i$ with $0 \leq r_i \leq m - 1$. However, $r_i = 0$ is impossible here since $(a, m) = 1$ and $m \nmid i$ (Theorem 2.12). Thus $1 \leq r_i \leq m - 1$. Thus $r_1, r_2, \ldots, r_{m-1}$ are all contained in $\{1, 2, \ldots, m - 1\}$. We now claim that they are distinct. Suppose that $r_i = r_j$. Then $q_i m + r_i \equiv q_j m + r_j \bmod m$. That is, $a \cdot i \equiv a \cdot j \bmod m$. Since $(a, m) = 1$, we have $i \equiv j \bmod m$. However, i and j both come from $1, 2, \ldots, m - 1$ and so by Lemma 4.9, $i = j$. That is, if $r_i = r_j$, then $i = j$. Thus $r_1, r_2, \ldots, r_{m-1}$ are distinct.

We have already shown that $\{r_1, \ldots, r_{m-1}\} \subset \{1, 2, \ldots, m - 1\}$. We now know that both of these sets contain exactly $m - 1$ distinct integers. Obviously, then, they must be equal. Since $\{r_1, \ldots, r_{m-1}\} = \{1, 2, \ldots, m - 1\}$, we have some rearrangement of $r_1, \ldots, r_{m-1}$, equaling (in order) $1, 2, \ldots, m - 1$. Because $i \cdot a = q_i m + r_i$, we have $i \cdot a \equiv r_i \bmod m$. Thus if $t_1, \ldots, t_{m-1}$ is the same rearrangement of $1 \cdot a, 2 \cdot a, \ldots, (m - 1) \cdot a$, we get $t_j \equiv j \bmod m$ and we are done. ∎

At the beginning of Chapter 2 we told you that for any integer a, $3 \mid a^3 - a$, that is, $a^3 \equiv a \bmod 3$. We are now ready to prove a general fact which has this statement as a corollary, namely that the statement is true for any integer a and any prime number p, not just 3. We will do this in two steps, the first step being the heart of the matter.

Theorem 4.11 (Fermat's Little Theorem). If p is a prime and $a \in \mathbb{Z}$ with $(a, p) = 1$, then $a^{p-1} \equiv 1 \bmod p$.

Proof. By Lemma 4.10, we have

$$\{a \cdot 1, a \cdot 2, \ldots, a(p - 1)\} \equiv \{1, 2, \ldots, p - 1\} \bmod p$$

Thus by Lemma 4.8,

$$(a \cdot 1)(a \cdot 2) \cdots (a(p - 1)) \equiv 1 \cdot 2 \cdots (p - 1) \bmod p$$

On the left-hand side, a occurs as a factor $p - 1$ times, so the congruence becomes

$$a^{p-1}(1 \cdot 2 \cdots (p - 1)) \equiv 1 \cdot 2 \cdots (p - 1) \bmod p$$

Now $(i, p) = 1$ for all $1 \leq i \leq p - 1$, and it follows from Exercise 3.A.23 that $(1 \cdot 2 \cdots (p-1), p) = 1$. Thus Theorem 4.2 permits us to cancel this product from both sides, leaving

$$a^{p-1} \equiv 1 \bmod p \qquad \blacksquare$$

Theorem 4.12 (Fermat's Little Theorem: second version). Let p be a prime number. If a is any integer, then

$$a^p \equiv a \bmod p$$

Proof. If $(a, p) = 1$, then by Theorem 4.6,

$$a^{p-1} \equiv 1 \bmod p$$

Hence multiplying by a yields

$$a^p \equiv a \bmod p$$

If, on the other hand, $(a, p) \neq 1$, then $(a, p) = p$, so $p \mid a$. Thus $a \equiv 0 \bmod p$ and so $a^p \equiv 0 \bmod p$ also. Thus $a^p \equiv 0 \equiv a \bmod p$. $\blacksquare$

You may be curious about the name attached to Theorem 4.11. Pierre Fermat (1601?–1665) is generally considered to have been the greatest theoretical mathematician of the seventeenth century. (Newton was an applied mathematician.) Particularly remarkable is that Fermat earned his living as a lawyer; more precisely, he was a King's Councillor and only did mathematics as a hobby! Fermat, obviously, was a genius, who seldom missed an opportunity (mathematically speaking). Theorem 4.11 represents one of the few opportunities that he did miss, for the Swiss mathematician Leonhard Euler (1707–1783) used the essence of the foregoing arguments more cleverly and found a better theorem, which is the topic of the next section. Thus Theorem 4.11 is "little," but only by comparison.

4.4 Euler's Theorem

Let us approach Euler's generalization of the above by looking at a skeletal outline of the proof of Theorem 4.11. In that proof we had a number $m(= p$, a prime) and a set of numbers $\{1, 2, \ldots, m - 1\}$ satisfying two conditions:

1. Every number in the set was relatively prime to $m(= p)$.
2. When each number in the set was multiplied by a, the resulting set $\{a \cdot 1, a \cdot 2, \ldots, a(m-1)\}$ was congruent modulo m to the original set.

We now see that the following theorem should be true.

Theorem 4.13. Let $m > 0$. If there are $n > 0$ distinct integers $c_1, \ldots, c_n$ each of which is relatively prime to m and an integer a such that

$$\{ac_1, ac_2, \ldots, ac_n\} \equiv \{c_1, c_2, \ldots, c_n\} \bmod m$$

then $a^n \equiv 1 \bmod m$.

Proof. By Lemma 4.8, $c_1c_2 \cdots c_n \equiv (ac_1)(ac_2) \cdots (ac_n) \bmod m$, hence $a^n(c_1c_2 \cdots c_n) \equiv c_1c_2 \cdots c_n \bmod m$. We have $(c_i, m) = 1$ for each $1 \leq i \leq n$, so $(c_1c_2 \cdots c_n, m) = 1$. Thus the factor $c_1c_2 \cdots c_n$ can be canceled and we are left with $a^n \equiv 1 \bmod m$. ▮

Theorem 4.11 is a special case of 4.13, for we see that if $m = p$ is a prime number and if $n = p - 1$ and if $\{c_1, \ldots, c_n\} = \{1, 2, \ldots, p - 1\}$, then the hypothesis of Theorem 4.13 is satisfied for any $a \in \mathbb{Z}$ relatively prime to p.

While it is true that 4.13 generalizes 4.11, it really is not too aesthetically pleasing, for its hypothesis is too vague. For instance, it refers to some integer $n > 0$ and some integer a. What we seek is something more specific, and this brings us to Euler's way of satisfying the hypothesis and at the same time being quite precise. In Theorem 4.11 we are told definitely to take $n = p - 1$ and a can be anything subject only to the condition that $(a, p) = 1$. Thus the beauty of Euler's work, which we are about to present, is that it is more general than Fermat's and yet definitive enough to be of interest. The true value of Theorem 4.13, then, is to reveal the relationship between the proof of Theorem 4.11 and what is to follow.

We proceed to an easy lemma.

Lemma 4.14. If $m > 0$ and if $u \equiv v \bmod m$, then $(u, m) = 1$ if and only if $(v, m) = 1$.

Proof. Suppose that $(u, m) = 1$. Now $m \mid u - v$, so $u - v = km$ for some $k \in \mathbb{Z}$ and so $u = v + km$. If $d = (v, m)$, then $d \mid v$ and $d \mid m$; hence $d \mid u$. Thus d is a common divisor of u and m, so $d = 1$ since $(u, m) = 1$. The converse statement is obtained by interchanging u and v. ▮

One significant factor in the proof of Theorem 4.11 was that if $m = p$ is prime, then every positive integer less than m is relatively prime to m. The key to Euler's work is that we may drop the requirement that m be a prime if we consider only those positive integers less than m which are relatively prime to m. We define a symbol φ known as *Euler's φ-function* (phi-function).

Definition. If $m > 1$, let $\varphi(m)$ denote the number of positive integers which are less than m and relatively prime to m. That is, $\varphi(m)$ is the number of elements in the set $\{c \in \mathbb{Z} \mid 1 \le c \le m - 1 \text{ and } (c, m) = 1\}$. For convenience we let $\varphi(1) = 1$.

Here are a few values of $\varphi(m)$. If $m = 2$, then $\varphi(m) = 1$, since 1 is the only positive integer less than 2 and $(1, 2) = 1$. If $m = 6$, then $\varphi(m) = 2$ since 1 and 5 are relatively prime to 6 while 2, 3, 4 are not. If $m = 12$, then $\{c \in \mathbb{Z} \mid 1 \le c \le 11 \text{ and } (c, 12) = 1\} = \{1, 5, 7, 11\}$, so $\varphi(12) = 4$.

Should you be wondering if there is a formula for calculating $\varphi(m)$ for any $m > 1$, we assure you that there is, but we will present it later. For the time being we will proceed along theoretical lines and hopefully demonstrate that while formulas at times are indispensable, much fine mathematics can be carried out without them. Thus understanding what $\varphi(m)$ is will carry us a long way. There is, of course, one case in which $\varphi(m)$ is quite easy to calculate. If $m = p$ is a prime number, then each integer $1 \le c \le p - 1$ is relatively prime to p, so $\varphi(p) = p - 1$. Keep this in mind when considering Theorem 4.15 and you will see that 4.11 is a special case of Theorem 4.15.

The definition of $\varphi(m)$ tells us that $\varphi(m)$ is the number of elements in the set $\{c \in \mathbb{Z} \mid 1 \le c \le m - 1 \text{ and } (c, m) = 1\}$. Thus we can number the members in this set by the integers $1, 2, 3, \ldots, \varphi(m)$. Hence we will write $\{c \in \mathbb{Z} \mid 1 \le c \le m - 1 \text{ and } (c, m) = 1\} = \{c_1, c_2, \ldots, c_{\varphi(m)}\}$.

Theorem 4.15 (Euler's Theorem). If $m > 0$ and $(a, m) = 1$, then

$$a^{\varphi(m)} \equiv 1 \bmod m$$

Proof. We will show that the hypothesis of Theorem 4.13 is satisfied by $n = \varphi(m)$, a, and $\{c_1, c_2, \ldots, c_{\varphi(m)}\} = \{c \in \mathbb{Z} \mid 1 \le c \le m - 1 \text{ and } (c, m) = 1\}$. Hence we must show that

$$\{ac_1, ac_2, \ldots, ac_{\varphi(m)}\} \equiv \{c_1, c_2, \ldots, c_{\varphi(m)}\} \bmod m$$

We proceed as in the proof of Fermat's Little Theorem. Let i be an integer, $1 \le i \le \varphi(m)$. Apply the Division Algorithm to ac_i and m to obtain integers q_i and r_i such that

$$ac_i = mq_i + r_i, \qquad 0 \le r_i < m$$

Since $(a, m) = 1$ and $(c_i, m) = 1$, we have $(ac_i, m) = 1$. Now $ac_i \equiv r_i \bmod m$ and so by Lemma 4.14, $(r_i, m) = 1$. In particular, $r_i \ne 0$, so that we now know that r_i lies between 1 and $m - 1$. Combining this with $(r_i, m) = 1$, we conclude that the set of remainders obtained $\{r_1, r_2, \ldots, r_{\varphi(m)}\} \subseteq \{c_1, c_2, \ldots, c_{\varphi(m)}\}$. We claim that there is no duplication among the remainders. Put another way, we claim that if $r_i = r_j$, then $i = j$. For if $r_i = r_j$, then

$$ac_i - mq_i = r_i = r_j = ac_j - mq_j$$

Hence $a(c_i - c_j) = m(q_i - q_j)$. Thus $m \mid a(c_i - c_j)$, but since $(a, m) = 1$ this leads to $m \mid c_i - c_j$ and hence $c_i \equiv c_j \bmod m$. However, both c_i and c_j lie between 1 and $m - 1$. By Lemma 4.9, $c_i = c_j$. As we were assuming that the c_i's were distinct, we must have $i = j$.

We already have $\{r_1, r_2, \ldots, r_{\varphi(m)}\} \subseteq \{c_1, c_2, \ldots, c_{\varphi(m)}\}$. However, we now see that both these sets consist of $\varphi(m)$ distinct elements, so $\{r_1, r_2, \ldots, r_{\varphi(m)}\} = \{c_1, c_2, \ldots, c_{\varphi(m)}\}$. Thus $r_1, r_2, \ldots, r_{\varphi(m)}$ can be rearranged to equal (in order) $c_1, c_2, \ldots, c_{\varphi(m)}$. Since $ac_i \equiv r_i \bmod m$, it follows from the definition that $\{ac_1, ac_2, \ldots, ac_{\varphi(m)}\} \equiv \{c_1, c_2, \ldots, c_{\varphi(m)}\} \bmod m$. Applying Theorem 4.13, we reach our goal,

$$a^{\varphi(m)} \equiv 1 \bmod m \qquad \blacksquare$$

Suppose that $m > 0, (a, m) = 1$ and let d be the order of a modulo m. Comparing Theorem 4.6 to Theorem 4.15, we get the significant result that $d \mid \varphi(m)$. We close this section with a question. If $d \mid \varphi(m)$, is there an a (with $(a, m) = 1$) such that d is the order of a? Not always (Exercise IV.A.25), but for some m it is true. More about these in Chapter 8.

4.5 A Historical View

If Fermat was as brilliant as he is reputed to have been, why did he only get as far as Theorem 4.11 and fail to discover Theorem 4.15? We can only speculate on the answer, but the following guess appears plausible. Fermat seldom published his work and often the only records of what he did are statements of theorems he discovered, usually given without proofs, in letters to his friends, in notebooks, in margins of books that he read, and so on. Theorem 4.11 was one such case. He proved it but his proof was never found, only the statement. Therefore, we do not know if the proof of Theorem 4.11 which we have given is the one that Fermat used. We think rather not, for we feel that had he used the foregoing proof he would indeed have pursued the matter further and uncovered Euler's Theorem. It is our guess that Fermat had a different proof and we would like to present our candidate. Let us point out that Fermat is credited with inventing a technique known as "infinite descent," and that this technique is essentially none other than our old friend from Chapter 1, mathematical induction. Our candidate for a proof uses induction. In addition, the argument also depends on the well-known formula for finding the number of ways in which subsets with k elements can be selected from a set with n elements. This formula is given by the number $\dfrac{n!}{k!(n-k)!}$ and denoted $\dbinom{n}{k}$. Here $m! = m(m-1)(m-2)\cdots 3 \cdot 2 \cdot 1$ for any integer $m \geq 1$, and by convention $0! = 1$. Fermat and Blaise Pascal (1623–1662) did much work involving this idea. Thus all of the ideas in our argument were well within Fermat's realm of knowledge. As far as we know, the proof cannot be generalized to give Theorem 4.15; this would explain, perhaps, why Fermat succeeded in discovering his "little" theorem but not Euler's.

The number $\binom{n}{k}$ occurs as the coefficient of $a^{n-k}b^k$ in the expansion of $(a + b)^n$. Thus we have the identity

$$(a + b)^n = a^n + \binom{n}{1}a^{n-1}b + \binom{n}{2}a^{n-2}b + \cdots + \binom{n}{k}a^{n-k}b^k + \cdots + \binom{n}{n-1}ab^{n-1} + b^n$$

We will assume that you are familiar with this expansion.

Lemma 4.16. If p is a prime number and if $1 < k < p$, then $p \,\Big|\, \binom{p}{k}$.

Proof. Suppose that $m = \binom{p}{k} = \dfrac{p!}{k!(p-k)!}$. Hence $p! = mk!(p-k)!$. Now $p \mid p!$ and so $p \mid mk!(p-k)!$; hence by Corollary 3.5, p divides one of m, $k!$ or $(p-k)!$. However, $1 < k < p$ and since $k! = k(k-1)(k-2)\cdots 3\cdot 2\cdot 1$, p cannot divide $k!$. Similarly, $1 < p - k < p$, so that p does not divide $(p-k)!$ either. We must then have $p \mid m$. ∎

Corollary 4.17. If p is a prime number and $a, b \in \mathbb{Z}$, then $p \mid (a + b)^p - a^p - b^p$.

Proof. Since

$$(a + b)^p - a^p - b^p = \binom{p}{1}a^{p-1}b + \binom{p}{2}a^{p-2}b^2 + \cdots + \binom{p}{p-2}a^2b^{p-2} + \binom{p}{p-1}ab^{p-1},$$

we apply Lemma 4.16 and Corollary 2.3. ∎

Fermat's Little Theorem Revisited. If p is a prime, then for any integer a,

$$a^p \equiv a \bmod p$$

Proof. We first work with $a \geq 0$. If $a = 0$ or $a = 1$, then $a^p = a$, so that $a^p \equiv a \bmod p$. For $a > 1$, we use induction. Thus suppose that $a \geq 1$ and

$a^p \equiv a \bmod p$. We must show that $(a+1)^p \equiv a + 1 \bmod p$. Now

$$(a+1)^p - (a+1) = [(a+1)^p - a^p - 1^p] + [a^p - a]$$

By Corollary 4.17, $p \mid (a+1)^p - a^p - 1^p$, and by induction, $p \mid a^p - a$. Thus $p \mid (a+1)^p - (a+1)$, as desired.

We leave the case $a < 0$ to the reader. For p an odd prime, use $(-a)^p = -a^p$, and for $p = 2$, the whole problem is trivial. ∎

4.6 Wilson's Theorem

We will now consider a result that uses the idea of a multiplicative inverse. A lemma precedes it.

Lemma 4.18. Let p be a prime number. If $a \in \mathbb{Z}$ with $1 < a < p - 1$ and b is the multiplicative inverse of a modulo p, then $a \neq b$.

Proof. First notice that a has a multiplicative inverse modulo p because $(a, p) = 1$. Now suppose that $b = a$. Then from $ab \equiv 1 \bmod p$, we have $a^2 \equiv 1 \bmod p$. Thus $p \mid a^2 - 1$. But $a^2 - 1 = (a-1)(a+1)$ and since p divides this product, it must divide one of the factors. Thus $p \mid a - 1$ or $p \mid a + 1$. If $p \mid a - 1$, then $a \equiv 1 \bmod p$, so $a = 1$ by Lemma 4.9; we have a contradiction in this case. If $p \mid a + 1$, then $a \equiv -1 \bmod p$. However $p - 1 \equiv -1 \bmod p$, so $a \equiv (p-1) \bmod p$, and again by Lemma 4.9, $a = p - 1$, which is also a contradiction. Since the assumption that $b = a$ leads to a contradiction, we must have $b \neq a$. ∎

It is now clear from the lemma that for a prime number p, the numbers $2, 3, 4, \ldots, p - 2$ can be paired off, each pair consisting of an integer and its multiplicative inverse modulo p. Thus for $p = 11$, we have the pairs $(2, 6)$, $(3, 4)$, $(5, 9)$, and $(7, 8)$. Suppose that we examine the product $(p-1)! = 10! = 1 \cdot 2 \cdot 3 \cdot 4 \cdot 5 \cdot 6 \cdot 7 \cdot 8 \cdot 9 \cdot 10$. We rearrange this product as $10! = 10(2 \cdot 6)(3 \cdot 4)(5 \cdot 9)(7 \cdot 8)$. Since $2 \cdot 6 \equiv 1 \bmod 11$, $3 \cdot 4 = 1 \bmod 11$, $5 \cdot 9 \equiv 1 \bmod 11$, and $7 \cdot 8 \equiv 1 \bmod 11$, we have $10! \equiv 10 \bmod 11$, so $10! + 1 \equiv 0 \bmod p$. This argument generalizes quite easily to what is known as *Wilson's Theorem.*

Theorem 4.19 (Wilson's Theorem). For an integer $p > 1$, p is a prime number if and only if $(p-1)! + 1 \equiv 0 \bmod p$.

Proof. Suppose first that p is a prime number. If $p = 2$, then $(p-1)! = 1! = 1$ and $1 + 1 \equiv 0 \bmod 2$. Hence assume that $p > 2$. In the product $(p-1)! = 2 \cdot 3 \cdot 4 \cdots (p-2)(p-1)$, we rearrange the integers in pairs, where each number between 2 and $p - 2$ is matched up with its multiplicative inverse modulo p. Thus $2 \cdot 3 \cdots (p - 2) \equiv 1 \bmod p$, since the product of a number and its multiplicative inverse is congruent to 1 mod p. Multiplying by $p - 1$, we get

$$(p - 1)! = 2 \cdot 3 \cdots (p - 2)(p - 1) \equiv p - 1 \bmod p$$

However, $p - 1 \equiv -1 \bmod p$, so

$$(p - 1)! \equiv -1 \bmod p$$
$$(p - 1)! + 1 \equiv 0 \bmod p$$

Conversely, suppose that $(p - 1)! + 1 \equiv 0 \bmod p$. We will show that p must be a prime number. If p is not prime, then p has a divisor k where $1 < k < p$. Thus k occurs as one of the terms in the product $(p - 1)!$ so $k \mid (p - 1)!$. Since $(p - 1)! + 1 \equiv 0 \bmod p$, we have $p \mid (p - 1)! + 1$, and since $k \mid p$, we conclude that $k \mid (p - 1)! + 1$. But $1 = [(p - 1)! + 1] - [(p - 1)!]$ and k divides each term in brackets, so $k \mid 1$. This is clearly impossible because $k > 1$. Accordingly, p is a prime number. ▮

Wilson's Theorem is J. Wilson's (1741–1793) one mark on history. Apparently, he did not even prove it, but only guessed that it was true. In fact, LaGrange was the first to prove it, although the proof just given is a nicer one due to Carl Friedrich Gauss (1777–1855). Many people consider Gauss to be the finest mathematician to ever live.

Wilson's Theorem is not very practical for finding primes. To see if 3,472,811 is prime, we would have to compute 3,472,810!, an awesome task. Yet Wilson's Theorem has great theoretical value. It helps prove our next result, which is crucial to work in Chapter 7.

Theorem 4.20. Let p be a prime number. Then there is a solution of

$$X^2 \equiv -1 \bmod p$$

if and only if $p \not\equiv 3 \bmod 4$.

Proof. If $p = 2$, then $1^2 \equiv 1 \equiv -1 \bmod p$. If $p \neq 2$, then since p is prime, we have either $p \equiv 1 \bmod 4$ or $p \equiv 3 \bmod 4$. Suppose first that $p \equiv 3 \bmod 4$. We must show that $X^2 \equiv -1 \bmod p$ has no solution. Assume that there is an integer x such that $x^2 \equiv 1 \bmod p$. Now $p = 4n + 3$ for some $n \in Z$, so that $(p - 1)/2 = 2n + 1$. This is odd. Thus $(-1)^{(p-1)/2} = -1$. However, since $x^2 \equiv -1 \bmod p$, we get

$$-1 = (-1)^{(p-1)/2} \equiv (x^2)^{(p-1)/2} = x^{p-1} \equiv 1 \bmod p$$

The last congruence is due to Fermat's Little Theorem. However, since $p \neq 2$, $-1 \not\equiv 1 \bmod p$, and we have a contradiction. Thus if $p \equiv 3 \bmod 4$, it is impossible to have $x^2 \equiv -1 \bmod p$.

We now suppose that $p \equiv 1 \bmod 4$. We will show that $X^2 \equiv -1 \bmod p$ does have a solution. In fact, we claim that if $x = 1 \cdot 2 \cdot 3 \cdots (p-1)/2$, then x is a solution. First notice that since $p = 4n + 1$ for some $n \in Z$, $(p-1)/2 = 2n$ is even. Now

$$\begin{aligned} x^2 &= [1 \cdot 2 \cdot 3 \cdots (p-1)/2] \cdot [1 \cdot 2 \cdot 3 \cdots (p-1)/2] \\ &= [1 \cdot 2 \cdot 3 \cdots (p-1)/2] \cdot [(-1)(-2)(-3) \cdots (-(p-1)/2)] \end{aligned}$$

since in the last expression we have introduced $(p-1)/2$ minus signs, and $(p-1)/2$ is even. Now since $-a \equiv p - a \bmod p$, the equation above gives

$$x^2 \equiv 1 \cdot 2 \cdot 3 \cdots (p-1)/2[(p-1)(p-2)(p-3) \cdots (p+1)/2] \bmod p$$

Here we have used the fact that $p - (p-1)/2 = (p+1)/2$. We now reverse the order in the last half of the foregoing product and get

$$x^2 \equiv [1 \cdot 2 \cdot 3 \cdots (p-1)/2][(p+1)/2 \cdots (p-3)(p-2)(p-1)] \bmod p$$

That is,

$$x^2 \equiv (p-1)! \bmod p$$

By Wilson's Theorem, $(p-1)! \equiv -1 \bmod p$, so $x^2 \equiv -1 \bmod p$, as desired. ∎

EXERCISES 4.A

1. Determine whether or not the following statements are true:
   (a) $7 \equiv 3 \bmod 4$ (b) $7 \equiv -3 \bmod 4$ (c) $7 \equiv 7 \bmod 4$
   (d) $325 \equiv -1 \bmod 37$ (e) $325 \equiv -8 \bmod 37$ (f) $-42 \equiv -11 \bmod 16$
2. Find a number x, $0 \leq x \leq 10$, such that $k^3 + 6k^2 - 24 \equiv x \bmod 11$ for each of the following values of k: 1, 12, 3, -8, -5, 6.
3. Verify that $1 \equiv 12 \bmod 11$, $3 \equiv -8 \bmod 11$, and $-5 \equiv 6 \bmod 11$. How is this related to your answers in Exercise 4.A.2?
4. Given that $k \equiv 16 \bmod 11$, find a number x, $0 \leq x \leq 10$ such that $k^3 + 6k^2 - 24 \equiv x \bmod 11$.
5. Show that $a \equiv b \bmod 1$ for any a and b.
6. Show that if $a \equiv b \bmod m$ and n is a positive divisor of m, then $a \equiv b \bmod n$.
7. Prove: If $(a, 2) = 1$, then $a^2 \equiv 1 \bmod 4$.
8. Let p be a prime number. Show that if $a, b \in \mathbb{Z}$ such that $ab \equiv 0 \bmod p$, then either $a \equiv 0 \bmod p$ or $b \equiv 0 \bmod p$.

9. Let p be a prime number. Show that if $x \in \mathbb{Z}$ such that $x^2 \equiv 1 \bmod p$, then either $x \equiv 1 \bmod p$ or $x \equiv -1 \bmod p$.

10. Prove: If p is a prime number, $p \neq 2$, then either $p \equiv 1 \bmod 4$ or $p \equiv 3 \bmod 4$.

11. For each of the following pairs of integers a and m, verify that $(a, m) \neq 1$ and check directly that no integer between 0 and $m - 1$ can be a multiplicative inverse for a modulo m:

$$a = 4, \quad m = 6; \qquad a = 10, \quad m = 6; \qquad a = 6, \quad m = 8;$$
$$a = -2, \quad m = 8$$

12. Note that $4 \equiv 10 \bmod 6$ and $6 \equiv -2 \bmod 8$. How is this related to your work in Exercise 4.A.11?

13. For which of the following values of m does 6 have a multiplicative inverse modulo m: $m = 5, m = 6, m = 7, m = 8$? Find it, if it exists.

14. Solve the following congruences by first finding the multiplicative inverse of the coefficient of X modulo the given m.

(a) $6X \equiv 3 \bmod 7$ (b) $7X \equiv -4 \bmod 9$
(c) $4X \equiv 22 \bmod 15$ (d) $19X \equiv 22 \bmod 15$
(e) $44X \equiv 61 \bmod 7$ (f) $413X \equiv 58 \bmod 412$
(g) $12X \equiv -18 \bmod 39$

15. Calculate $5^0, 5^1, 5^2, \ldots$ modulo 13 until the cycle has repeated three times.

16. Find the order modulo 13 for each of the following integers: 2, 5, 11. What is the multiplicative inverse of each of these modulo 13?

17. Determine which of the following statements are true.

(a) $\{2, -3\} \equiv \{10, 5\} \bmod 8$
(b) $\{17, 8\} \equiv \{-3, 6\} \bmod 11$
(c) $\{6, 13\} \equiv \{-18, 42\} \bmod 24$
(d) $\{4, 6, 12\} \equiv \{-3, 14, 10\} \bmod 3$
(e) $\{19, -18, 70, -8, 22, 32, -348\} \equiv \{0, 1, 2, 3, 4, 5, 6\} \bmod 7$

18. Find a set $\{r_1, r_2, \ldots, r_7\}$ such that for each i, $0 \leq r_i \leq 9$ and $\{r_1, r_2, \ldots, r_7\} \equiv \{12, -35, -9, 46, -12, 20, 24\} \bmod 10$.

19. For $p = 7$, verify Fermat's Little Theorem for each of the following values of a: $0, 2, -3, -6, 3, 12, 6$.

20. Use Fermat's Little Theorem to solve for x in each of the following.

(a) $2^{11} \equiv x \bmod 3, 0 \leq x \leq 2$
(b) $2^{11} \equiv x \bmod 7, 0 \leq x \leq 6$
(c) $3^{44} \equiv x \bmod 7, 0 \leq x \leq 6$
(d) $11^{70} \equiv x \bmod 13, -25 \leq x \leq -13$

21. Evaluate: $\varphi(15)$, $\varphi(21)$, $\varphi(24)$, $\varphi(32)$.

22. Use Euler's Theorem to solve for x in each of the following.

(a) $2^{14} \equiv x \bmod 15, 0 \leq x \leq 14$

(b) $5^{17} \equiv x \bmod 21, 21 \leq x \leq 41$

(c) $(29)^{25} \equiv x \bmod 24, -23 \leq x \leq 0$

(d) $(-533)^{18} \equiv x \bmod 32, 160 \leq x \leq 191$

23. For $p = 13$, rearrange the terms in the product 12! by matching each integer with its multiplicative inverse modulo 13 and verify this special case of Wilson's Theorem.

24. Verify that $12 \nmid 11! + 1$. Does this violate Wilson's Theorem?

25. Find an $m > 0$ and $d > 0$ with $d \mid \varphi(m)$ such that d is not the order of any a modulo m, $(a, m) = 1$.

EXERCISES 4.B

1. If $a \equiv b \bmod m$ and $d \in \mathbb{Z}$ such that $a = Ad$, $b = Bd$, $m = Md$, then $A \equiv B \bmod M$.

2. Prove: If $ac \equiv bc \bmod m$ and $d = (c, m)$, then $a \equiv b \bmod (m/d)$.

3. Prove: If $a \equiv b \bmod m$, then $(a, m) = (b, m)$.

4. Show that if $a \equiv b \bmod p$ for all primes p, then $a = b$.

5. Show that if $(a, m) = 1$ and c is the multiplicative inverse of a modulo m, then the congruence $aX \equiv b \bmod m$ has for its solutions the set $\{x \mid x \equiv bc \bmod m\}$.

6. Let $p > 2$ be a prime. Show that if $(a, p) = 1$, then either $a^{(p-1)/2} \equiv 1 \bmod p$ or $a^{(p-1)/2} \equiv -1 \bmod p$.

7. Let p and q be distinct primes. Show that $p^{q-1} + q^{p-1} \equiv 1 \bmod pq$. (*Hint:* Fermat; note that $(p, q) = 1$.]

8. Prove: If m is not a prime and $m > 1$, then $\varphi(m) < m - 1$.

9. Let p be a prime; calculate $\varphi(p^2)$.

10. Show that $\binom{n}{0} - \binom{n}{1} + \binom{n}{2} \cdots \pm \binom{n}{n} = 0$ [*Hint:* Find A and B such that $(A - B)^n$ yields this expression.]

11. Let $p \neq 2$ be a prime, and let $x = \left(\frac{p-1}{2}\right)!$ In the proof of Theorem 4.20, we saw that if $p \equiv 1 \bmod 4$, then $x^2 \equiv -1 \bmod p$. Prove that if $p \equiv 3 \bmod 4$, then $x^2 \equiv 1 \bmod p$.

12. (a) Show that if an integer x is the sum of two squares, then $x \not\equiv 3 \bmod 4$.

(b) Show that if an integer x is the sum of three squares, then $x \not\equiv 7 \bmod 8$.

13. Find all integers a, $0 < a < 13$, such that $X^2 \equiv a \bmod 13$ has a solution. How many such a are there? Repeat modulo 17, 19, and 23. Guess a theorem.

14. (a) Show that the equation $3x + 2 = y^2$ has no integer solutions.
(b) Show that the equation $5x + 3 = y^2$ has no integer solutions.

15. Let $m > 0$ and let $\{c_1, c_2, \ldots, c_{\varphi(m)}\} = \{c \mid 1 \le c \le m \text{ and } (c, m) = 1\}$. Show that $c_1 + c_2 + \cdots + c_{\varphi(m)} \equiv 0 \bmod m$.

16. (a) If $p > 2$ is prime and $a \in Z$, show that $X^2 \equiv a \bmod p$ has either no solutions or exactly two solutions modulo p (also see Exercise 4.A.9).
(b) If this congruence has two solutions, show that only one solution, x, satisfies $0 \le x \le (p - 1)/2$. (*Hint*: After you find one, what *must* the other one be?)

5

Multiplicative Functions

In Chapter 4 we introduced you to Euler's φ-function and gave you an instance of how it played an important role in number theory. We will see in Chapter 8 another of its uses. This chapter is devoted to developing properties of φ. In doing so, we will see that there are other functions which have similar characteristics and we will examine these.

5.1 Three Special Functions

Both Euler's Theorem and material yet to come suggest that it would be quite useful to have available a concise and working formula for $\varphi(m)$; that is, we would prefer something that would free us from having to enumerate the numbers between 1 and m, isolate those relatively prime to m, and count how many of these there are. We start with the key result. Namely, we will show that if n and m are relatively prime, then $\varphi(nm) = \varphi(n)\varphi(m)$. Since the argument involves counting things in a certain manner, let us first look at an example with $n = 35$ and $m = 6$.

We wish to count the number of integers between 1 and $nm = 210$ which are relatively prime to 210 and see that this number is exactly $\varphi(35) \cdot \varphi(6)$. Note that since $(35, 6) = 1$, we must have $(a, 210) = 1$ if and only if $(a, 35) = 1$ and $(a, 6) = 1$. Thus we wish to count the number of integers between 1 and 210 which are relatively prime to both 35 and 6.

We begin by arranging the numbers from 1 to 210 in the following array, displayed so as to show the congruence classes modulo 6:

$$\begin{array}{cccccc}
0\cdot 6+1 & 0\cdot 6+2 & 0\cdot 6+3 & 0\cdot 6+4 & 0\cdot 6+5 & 0\cdot 6+6 \\
1\cdot 6+1 & 1\cdot 6+2 & 1\cdot 6+3 & 1\cdot 6+4 & 1\cdot 6+5 & 1\cdot 6+6 \\
2\cdot 6+1 & 2\cdot 6+2 & 2\cdot 6+3 & 2\cdot 6+4 & 2\cdot 6+5 & 2\cdot 6+6 \\
\vdots & \vdots & \vdots & \vdots & \vdots & \vdots \\
34\cdot 6+1 & 34\cdot 6+2 & 34\cdot 6+3 & 34\cdot 6+4 & 34\cdot 6+5 & 34\cdot 6+6
\end{array}$$

(Note that the last entry is $35 \cdot 6 = 210$.) We will first find those numbers which are relatively prime to 6 and among those find those numbers which are relatively prime to 35.

Examine the fourth column; its entries consist of numbers of the form $k \cdot 6 + 4$, where $0 \leq k \leq 34$. Since $(4, 6) = 2$, clearly $(k \cdot 6 + 4, 6) = 2 \neq 1$. Thus we may eliminate all numbers in this column from consideration. The same applies to the second, third, and sixth columns; none of these columns contains numbers relatively prime to 6. We are left with columns one and five. Since $(5, 6) = 1$ then $(k \cdot 6 + 5, 6) = 1$ and similarly $(k \cdot 6 + 1, 6) = 1$. We have now found all the numbers between 1 and 210 that are relatively prime to 6; they are the numbers in either column one or column five. Note that this is because $(1, 6) = (5, 6) = 1$. Also, the number of such columns is precisely $\varphi(6) = 2$.

Next, we have to eliminate from columns one and five those numbers not relatively prime to 35; the numbers remaining will then be relatively prime to both 35 and 6 and hence relatively prime to $210 = 35 \cdot 6$. Look at the fifth column. The entries here are $0 \cdot 6 + 5$, $1 \cdot 6 + 5$, $2 \cdot 6 + 5, \ldots, 33 \cdot 6 + 5$, $34 \cdot 6 + 5$. Recalling a concept from Chapter 4, we claim that

$$\{0 \cdot 6 + 5, 1 \cdot 6 + 5, 2 \cdot 6 + 5, \ldots, 34 \cdot 6 + 5\} \equiv \{1, 2, 3, \ldots, 35\} \bmod 35$$

This is true for the following reasons. By Lemma 4.10,

$$\{1 \cdot 6, 2 \cdot 6, \ldots, 34 \cdot 6\} \equiv \{1, 2, \ldots, 34\} \bmod 35$$

because $(6, 35) = 1$. Now $0 = 0 \cdot 6$, so throwing in 0 we get

$$\{0 \cdot 6, 1 \cdot 6, 2 \cdot 6, \ldots, 34 \cdot 6\} \equiv \{0, 1, 2, \ldots, 34\} \bmod 35$$

Next add 5 to everything, which gives

$$\{0 \cdot 6 + 5, 1 \cdot 6 + 5, 2 \cdot 6 + 5, \ldots, 34 \cdot 6 + 5\} \equiv \{5, 6, 7, \ldots, 39\} \bmod 35$$

Finally, it is clear that

$$\{5, 6, 7, \ldots, 39\} \equiv \{1, 2, 3, \ldots, 35\} \bmod 35$$

so

$$\{0 \cdot 6 + 5, 1 \cdot 6 + 5, 2 \cdot 6 + 5, \ldots, 34 \cdot 6 + 5\} \equiv \{1, 2, 3, \ldots, 35\} \bmod 35$$

as claimed.

What all this shows is that every entry in the fifth column is congruent modulo 35 to one and only one member of the set $\{1, 2, \ldots, 35\}$. Hence we can work with this set instead. The numbers in this set which are relatively prime to 35 correspond to the numbers in the fifth column which are relatively prime to 35, and there are exactly $\varphi(35)$ of these. Thus the fifth column yields $\varphi(35)$ numbers relatively prime to both 35 and 6 (and hence to 210). Repeating the argument, the first column provides another $\varphi(35)$ numbers. As there were

$2 = \varphi(6)$ columns being dealt with, each with $\varphi(35)$ numbers relatively prime to 210, we arrive at the conclusion $\varphi(210) = \varphi(6)\varphi(35)$.

Keeping in mind this example, we proceed to the general proof.

Theorem 5.1. Let m and n be positive integers with $(m, n) = 1$. Then $\varphi(mn) = \varphi(m)\varphi(n)$.

Proof. We first remark that because $(n, m) = 1$, we have $(a, nm) = 1$ if and only if $(a, m) = 1$ and $(a, n) = 1$. Thus what we wish to count are the integers a satisfying the conditions $1 \leq a \leq nm$, $(a,n) = 1$, and $(a, m) = 1$.

Let a be a positive integer, $a \leq nm$. By a tiny variation of the Division Algorithm, we write $a = kn + r$, where $1 \leq r \leq n$; further, $0 \leq k \leq m - 1$. If $(r, n) \neq 1$, then $(a, n) \neq 1$; conversely, if $(r, n) = 1$, then $(a, n) = 1$. Thus the integers a which satisfy $(a, n) = 1$ are precisely those with representation $a = kn + r$, where $(r, n) = 1$. Now let $r_1, r_2, \ldots, r_{\varphi(n)}$ be the numbers in the set $\{1, 2, \ldots, n\}$ which are relatively prime to n. Then the numbers a lying between 1 and nm which are relatively prime to n are those of the form $a = kn + r_i$, where $i = 1, 2, \ldots, \varphi(n)$ and $1 \leq k \leq m - 1$.

Choose any i between 1 and $\varphi(n)$ and consider the numbers

$$0n + r_i,\ 1 \cdot n + r_i,\ 2n + r_i, \ldots, (m - 1)n + r_i$$

By Lemma 4.10, we have

$$\{n \cdot 1, n \cdot 2, \ldots, n(m - 1)\} \equiv \{1, 2, \ldots, m - 1\} \bmod m$$

Inserting $0 = n \cdot 0$ in each set and adding r_i to each number, we get

$$\{0n + r_i, 1n + r_i, 2n + r_i, \ldots, (m - 1)n + r_i\} \equiv \{r_i, r_i + 1, r_i + 2, \ldots, r_i + (m - 1)\} \bmod m$$

It is clear that

$$\{r_i, r_i + 1, r_i + 2, \ldots, r_i + (m - 1)\} \equiv \{1, 2, 3, \ldots, m\} \bmod m$$

so

$$\{0n + r_i, 1n + r_i, 2n + r_i, \ldots, (m - 1)n + r_i\} \equiv \{1, 2, 3, \ldots, m\} \bmod m$$

Thus the set

$$\{0n + r_i, 1n + r_i, 2n + r_i, \ldots, (m - 1)n + r_i\}$$

contains exactly $\varphi(m)$ numbers relatively prime to m and hence to mn. Since this statement holds for each of $r_1, r_2, \ldots, r_{\varphi(n)}$, we arrive at the result: there are exactly $\varphi(m) \cdot \varphi(n)$ numbers relatively prime to mn. Put another way, $\varphi(mn) = \varphi(m)\varphi(n)$. ∎

With this theorem in hand we can obtain a concise formula for $\varphi(m)$ which depends on the prime factorization of m. For if

$$m = p_1^{e_1}p_2^{e_2}\cdots p_k^{e_k}, \quad \text{then} \quad (p_i^{e_i}, p_j^{e_j}) = 1 \qquad \text{for } i \neq j$$

so $\varphi(m) = \varphi(p_1^{e_1})\varphi(p_2^{e_2})\cdots\varphi(p_k^{e_k})$. Thus we will be done once we know how to calculate φ of prime powers.

Lemma 5.2. Let p be a prime and $e \in \mathbb{Z}$ with $e \geq 1$. Then $\varphi(p^e) = p^{e-1}(p-1)$.

Proof. There are p^e elements between 1 and p^e and out of these we wish to count those which are relatively prime to p^e. We ask: How can an integer fail to be relatively prime to p^e? The answer is simply stated: $(a, p^e) \neq 1$ if and only if $p \mid a$. Thus to evaluate $\varphi(p^e)$ we count all the numbers from 1 to p^e (there are p^e of these) and subtract off the multiples of p between 1 and p^e. The multiples of p between 1 and p^e are $p, 2p, 3p, \ldots, (p^{e-1})p$, so there are exactly p^{e-1} integers between 1 and p^e which are not relatively prime to p^e. Therefore, $\varphi(p^e) = p^e - p^{e-1} = p^{e-1}(p-1)$. ∎

Theorem 5.3. Let $m \in \mathbb{Z}$ with $m \geq 2$. If the canonical representation of m is $m = p_1^{e_1}p_2^{e_2}\cdots p_k^{e_k}$, then $\varphi(m) = [p_1^{e_1-1}(p_1 - 1)][p_2^{e_2-1}(p_2 - 1)]\cdots[p_k^{e_k-1}(p_k - 1)]$.

Proof. By Theorem 5.1, $\varphi(m) = \varphi(p_1^{e_1})\varphi(p_2^{e_2})\cdots\varphi(p_k^{e_k})$ and by Lemma 5.2, $\varphi(p_i^{e_i}) = p_i^{e_i-1}(p_i - 1)$ for each $i = 1, \ldots, k$. ∎

When discussing functions such as φ, the following terminology is useful.

Definition.

(i) A function f defined on the positive integers and taking values in $\mathbb{Z}$ is called an *arithmetic function.*

(ii) An arithmetic function is called *multiplicative* if for any positive integers m and n with $(m, n) = 1$, we have $f(mn) = f(m)f(n)$.

For example, $f(n) = 2n$ is arithmetic but not multiplicative since $f(2)f(3) = 4 \cdot 6 = 24$, while $f(2 \cdot 3) = f(6) = 12$. $g(n) = n/2$ is not arithmetic since $g(3) = 3/2 \notin \mathbb{Z}$. Of course, φ is arithmetic, and Theorem 5.1 tells us that φ is multiplicative. The nice feature of a multiplicative function f is that if $n = p_1^{e_1}\cdots p_k^{e_k}$ is the prime decomposition of n, then $f(n) = f(p_1^{e_1})f(p_2^{e_2})\cdots f(p_k^{e_k})$. Hence, as we saw in Theorem 5.3, it is enough to know $f(p^e)$ for p any prime. We now look again at the function τ, introduced in Chapter 3, and a related function σ.

Definition. For any $n \geq 1$, let $\tau(n)$ denote the number of positive divisors of n and let $\sigma(n)$ denote the sum of the positive divisors of n.

From Corollary 3.8 we already have available a formula for $\tau(n)$, namely if $n \geq 2$, and $n = p_1^{e_1}p_2^{e_2}\cdots p_k^{e_k}$ is the canonical representation of n, then

$$\tau(n) = (e_1 + 1)(e_2 + 1)\cdots(e_k + 1)$$

Of course, $\tau(1) = 1$.

Theorem 5.4. The function τ is a multiplicative function.

Proof. Let m and n be positive integers with $(m, n) = 1$. We must show that $\tau(mn) = \tau(m)\tau(n)$. If either $m = 1$ or $n = 1$, the result is trivially true, so we assume that $m \geq 2$ and $n \geq 2$. Write out the canonical representations for m and n:

$$m = p_1^{e_1}p_2^{e_2}\cdots p_k^{e_k} \qquad \text{and} \qquad n = q_1^{f_1}q_2^{f_2}\cdots q_t^{f_t}$$

Because $(m, n) = 1$, the canonical representation for mn is (except for order)

$$mn = p_1^{e_1}p_2^{e_2}\cdots p_k^{e_k}q_1^{f_1}q_2^{f_2}\cdots q_t^{f_t}$$

Calculate $\tau(mn)$ from the formula to get

$$\tau(mn) = (e_1 + 1)(e_2 + 1)\cdots(e_k + 1)(f_1 + 1)(f_2 + 1)\cdots(f_t + 1)$$

Now calculate $\tau(m) = (e_1 + 1)(e_2 + 1)\cdots(e_k + 1$ and $\tau(n) = (f_1 + 1)\cdot(f_2 + 1)\cdots(f_t + 1)$. Compare $\tau(mn) = \tau(m)\tau(n)$. ∎

At this point, the function τ is relatively uninteresting, but its cousin σ has a bit more character. As you might expect, σ turns out to be a multiplicative function. The heart of that argument is given by the next lemma, which we will use again later.

Lemma 5.5. Let m and n be positive integers with $(m, n) = 1$. Let $a_1, a_2, \ldots, a_r$ be the distinct positive divisors of m, and let $b_1, b_2, \ldots, b_t$ be the distinct positive divisors of n. Then the distinct positive divisors of mn are just $a_1b_1, \ldots, a_1b_t, a_2b_1, \ldots, a_2b_t, \ldots, a_rb_1, \ldots, a_rb_t$.

Proof. First note that for $1 \leq i \leq r$ and $1 \leq j \leq t$, $a_i \mid m$ and $b_j \mid n$, so that a_ib_j is obviously a divisor of mn.

Second, we show that these products are all distinct. For suppose that $a_ib_j = a_kb_l$. Then $a_i \mid a_kb_l$. However, $(a_i, b_l) = 1$ [since $(m, n) = 1$ and $a_i \mid m$, $b_l \mid n$], so $a_i \mid a_k$. Now interchange the role of a_i and a_k, and by symmetry we see that $a_k \mid a_i$. Thus $a_i = a_k$. By cancellation, $b_j = b_l$. Thus from $a_ib_j = a_kb_l$ we have deduced that $a_i = a_k$ and $b_j = b_l$ so that the given list of products contains distinct elements.

Finally, since $a_1, \ldots, a_r$ are the distinct positive divisors of m, $\tau(m) = r$. Similarly, $\tau(n) = t$. By Theorem 5.4, $rt = \tau(m)\tau(n) = \tau(mn)$. That is, mn has rt distinct positive divisors. As our given list obviously contains rt divisors of mn, all of which are distinct, it must constitute the entire list of divisors of mn. ∎

Theorem 5.6. Let m and n be positive integers with $(m, n) = 1$. Then $\sigma(mn) = \sigma(m)\sigma(n)$.

Proof. If the positive divisors of m are $a_1, a_2, \ldots, a_r$ and the positive divisors of n are $b_1, b_2, \ldots, b_t$, then $\sigma(m) = a_1 + \cdots + a_r$ and $\sigma(n) = b_1 + \cdots + b_t$. By Lemma 5.5, the positive divisors of mn are $a_1b_1, \ldots, a_1b_t, \ldots, a_rb_1, \ldots, a_rb_t$, so that $\sigma(mn) = a_1b_1 + \cdots + a_1b_t + \cdots + a_rb_1 + \cdots + a_rb_t$. However, this equals $(a_1 + \cdots + a_r)(b_1 + \cdots + b_t) = \sigma(m)\sigma(n)$. ∎

As a corollary we obtain a formula for $\sigma(m)$ in terms of the prime factorization of m.

Corollary 5.7. If $m = p_1^{e_1}p_2^{e_2}\cdots p_k^{e_k}$, then

$$\sigma(m) = \left[\frac{p_1^{e_1+1} - 1}{p_1 - 1}\right]\left[\frac{p_2^{e_2+1} - 1}{p_2 - 1}\right]\cdots\left[\frac{p_k^{e_k+1} - 1}{p_k - 1}\right]$$

Proof. From the theorem,

$$\sigma(m) = \sigma(p_1^{e_1})\sigma(p_2^{e_2})\cdots\sigma(p_k^{e_k})$$

However, the positive divisors of $p_i^{e_i}$ are $1, p_i, p_i^2, p_i^3, \ldots, p_i^{e_i}$, so

$$\sigma(p_i^{e_i}) = 1 + p_i + p_i^2 + \cdots + p_i^{e_i} = \frac{p_i^{e_i+1} - 1}{p_i - 1}$$ ∎

Our interest in σ is more than its multiplicative property, as we shall see in the next section.

5.2 Perfect Numbers

Greeks of Euclid's time noticed early that the numbers 6 and 28 shared what they considered a mystical property: $6 = 1 + 2 + 3$ and $28 = 1 + 2 + 4 + 7 + 14$. That is, 6 and 28 are each the sum of their divisors, excluding themselves. There are other such numbers as we will see, but further discussion requires naming them.

Definition. An integer $n \geq 1$ is said to be a *perfect number* if $\sigma(n) = 2n$.

The name *perfect numbers* was provided by the Greeks; Euclid showed them how to construct some perfect numbers.

Theorem 5.8 (Euclid). Let $k \geq 1$ be an integer such that $2^k - 1$ is a prime number. Then $n = 2^{k-1}(2^k - 1)$ is a perfect number.

Proof. Because $2^k - 1$ is odd, we can apply Theorem 5.6 to get $\sigma(n) = \sigma(2^{k-1})\sigma(2^k - 1)$. Now $\sigma(2^{k-1}) = 2^{(k-1+1} - 1/2 - 1 = 2^k - 1$, so $\sigma(n) = (2^k - 1) \cdot \sigma(2^k - 1)$. As $2^k - 1$ is a prime, its only positive divisors are 1 and $2^k - 1$. Thus $\sigma(2^k - 1) = 1 + (2^k - 1) = 2^k$, which gives us

$$\sigma(n) = (2^k - 1) \cdot 2^k = 2[2^{k-1}(2^k - 1)] = 2n \qquad \blacksquare$$

From this theorem we see that $6 = 2 \cdot 3 = 2(2^2 - 1)$, $28 = 2^2(2^3 - 1)$, and $496 = 2^4(2^5 - 1)$ are perfect numbers. The difficulty in this theorem is finding prime numbers of the form $2^k - 1$. (See the discussion in Chapter 3.) Also, Euclid's theorem only produces even perfect numbers. The strength of Theorem 5.8 is that it does produce *every* even perfect number, as we now show. There is some evidence that the ancient Greeks may have known this as well. However, if so, their proof has not come down to us, and it was Euler who gave the following proof, 1800 years after Euclid.

Theorem 5.9 (Euler). If n is an even perfect number, then $n = 2^{k-1}(2^k - 1)$, where $2^k - 1$ is a prime number.

Proof. Let 2^{k-1} be the highest power of 2 dividing n. Then $k - 1 \geq 1$ and $n = 2^{k-1}m$, where $(2, m) = 1$. Because n is a perfect number, $\sigma(n) = 2n = 2(2^{k-1}m) = 2^k m$. On the other hand, $(2^{k-1}, m) = 1$, so $\sigma(n) = \sigma(2^{k-1})\sigma(m) = (2^k - 1)\sigma(m)$. Equating these two expressions for $\sigma(n)$, we obtain

$$2^k \cdot m = (2^k - 1)\sigma(m)$$

Hence $2^k - 1 \mid 2^k m$ and therefore $2^k - 1 \mid m$, because $(2^k - 1, 2^k) = 1$. We thus have $m = (2^k - 1)a$ for some $a \in \mathbb{Z}$, $a \geq 1$. Substituting this value of m, we then have

$$2^k(2^k - 1)a = (2^k - 1)\sigma(m)$$

so

$$\sigma(m) = 2^k \cdot a$$

However, $a \mid m$ and $m + a = (2^k - 1)a + a = 2^k a = \sigma(m)$. Since $\sigma(m)$ is the sum of all the divisors of m, we conclude that m and a are the only positive divisors of m. Furthermore, 1 is always a divisor of m and therefore a must be 1. Thus the only positive divisors of m are 1 and m, and hence m is a prime number. Finally, from $m = (2^k - 1)a$, we have $m = 2^k - 1$ and $n = 2^{k-1}m = 2^{k-1}(2^k - 1)$. $\blacksquare$

Hopefully, you wonder what can be said concerning odd perfect numbers. No one has ever found an example of an odd perfect number or a proof that there are none. Here then is another shadow cast by the great unknown.

An interesting sidelight of Euler's Theorem concerns primes p for which $2^p - 1$ is a prime. Primes that have the form $2^p - 1$ are called *Mersenne primes.* As we have mentioned, p must be a prime number whenever $2^p - 1$ is a prime. In 1644 a French priest, Father Marin Mersenne, asserted that the only primes $p \leq 257$ for which $2^p - 1$ is prime are $p = 2, 3, 5, 7, 13, 17, 19, 31, 67, 127, 257$; the name attached to these primes derives from this statement. It turned out that Mersenne was not completely accurate; neither $2^{67} - 1$ nor $2^{257} - 1$ are primes, while the primes $2^{61} - 1$, $2^{89} - 1$, and $2^{107} - 1$ were overlooked by him. We should not fault Mersenne for these inaccuracies since his work was done prior to the days of high-speed calculators. The fact that $2^{257} - 1$ is not prime was not established until 1932 by D. H. Lehmer (M. Kraitchik had announced this in 1922). At this time it is still not known if there are an infinite number of Mersenne primes (or, equivalently, if there are infinitely many even perfect numbers). The largest known Mersenne prime is the one mentioned at the beginning of Chapter 3, $2^{21701} - 1$. The difficulty of working with such primes is, of course, their huge size. The number $2^{21701} - 1$ has over 6500 digits.

Having considered primes of the form $2^k - 1$, what about primes of the form $2^k + 1$? As we saw in Exercise 3.B.10, if a prime p is of the form $2^k + 1$, then k must be a power of 2. In any discussion of these primes, there are at least three names that arise. The primes are called *Fermat primes* since Fermat guessed in 1640 that all numbers of the form $2^{2^k} + 1$ are prime numbers. Euler, however, showed (after some clever preliminary work) that

$$2^{2^5} + 1 = 4{,}294{,}967{,}297$$

was divisible by 641. This is the only known time Fermat was shown to have been wrong. However, even here he did not claim to have proved his statement; he clearly stated that he only believed it. In contrast, Fermat asserted, in a note written in the margin of a text, that he could prove that for every integer $n > 2$, the equation $x^n + y^n = z^n$ has no integral solution. A proof or a disproof of this statement, known as Fermat's last Theorem, has yet to be found by anyone.

Returning to Fermat primes, it was subsequently shown that $2^{64} + 1$, $2^{128} + 1$, and $2^{256} + 1$ are not primes. But perhaps the real interest in Fermat primes lies in a result of Gauss' which deals with the Euclidean construction of regular polygons having n sides. In Euclid's *Elements* a process is given for bisecting any angle using straightedge and compass. The problem of providing a Euclidean construction of an n-sided regular polygon is really that of providing a construction which divides the circumference of a circle into n equal segments. In turn, this amounts to constructing an angle whose measure is $360°/n$. Since angles can be bisected, the problem really involves those integers n which are odd. Gauss' brilliant result was that *for an odd integer n, a regular polygon having n sides can be constructed if and only if n is a product of distinct Fermat primes*! It was the unexpected beauty of this fact that made Gauss decide to dedicate his life to mathematics. He was 19.

5.3 Möbius Inversion

Having had a brief glimpse at what surely ranks as a highpoint of mathematical discovery, we return to a lower plane and discuss how to generate multiplicative functions. Suppose that f is an arithmetic function (and at this point we do not insist that it be multiplicative). We construct another arithmetic function in the following manner. For $n \geq 1$, let $d_1, d_2, \ldots, d_r$ be all of the positive divisors of n. Then define $F(n)$ to be $f(d_1) + f(d_2) + \cdots + f(d_r)$. In order to keep the notation short, we use *sigma notation* to abbreviate this sum. That is, $\sum_{d|n} f(d) = f(d_1) + f(d_2) + \cdots + f(d_r) = F(n)$. [Here we tolerate a bit of imprecision. Since we are looking at positive divisors of n, we should write $\sum_{\substack{d|n \\ d>0}} f(d)$, but for the sake of the typesetter we will make the the " $d > 0$ " invisible but tacitly understood.]

As an example, we point out that the function σ can be constructed in this way. For if we start with the identity function $f(n) = n$, and if the positive divisors of n are $d_1, d_2, \ldots, d_r$, then $\sum_{d|n} f(d) = f(d_1) + \cdots + f(d_r) = d_1 + \cdots + d_r = \sigma(n)$.

Our major goal is to show that if $F(n) = \sum_{d|n} f(d)$, then F is multiplicative if and only if f is multiplicative [note, for example, that σ is multiplicative, and obviously the identity function $f(n) = n$ is multiplicative]. We do half of this immediately, in a proof looking much like the proof of Theorem 5.6.

Theorem 5.10. Let f be a multiplicative function, and for $n \geq 1$ let $F(n) = \sum_{d|n} f(n)$. Then F is multiplicative.

Proof. Let m and n be positive integers with $(m, n) = 1$. Let $d_1, d_2, \ldots, d_r$ be all of the positive divisors of m, and $e_1, e_2, \ldots, e_t$ be all of the positive divisors of n. By Lemma 5.5 we know that $\{d_i e_j | 1 \leq i \leq r, 1 \leq j \leq t\}$ is the set of positive divisors of mn. Now $F(m) = \sum_{d|m} f(d) = f(d_1) + \cdots + f(d_r)$

and $F(n) = \sum_{e|n} f(e) = f(e_1) + \cdots + f(e_t)$. Multiplying these two expressions, we get

$$\begin{aligned} F(m)F(n) = f(d_1)[f(e_1) + f(e_2) + \cdots + f(e_t)] \\ + f(d_2)[f(e_1) + f(e_2) + \cdots + f(e_t)] + \cdots \\ + f(d_r)[f(e_1) + f(e_2) + \cdots + f(e_t)] \end{aligned}$$

Thus $F(m) \cdot F(n)$ is the sum of all products of the form $f(d_i)f(e_j)$, where $1 \le i \le r$, $1 \le j \le t$. Since $d_i | m$ and $e_j | n$, we have $(d_i, e_j) = 1$. Then because f is a multiplicative function $f(d_i)f(e_j) = f(d_i e_j)$. Thus $F(m)F(n)$ is the sum of all the numbers $f(d_i e_j)$, where $1 \le i \le r$, $1 \le j \le t$. But this is nothing more than $F(mn)$ since $\{d_i e_j | 1 \le i \le r, 1 \le j \le t\}$ is the set of all positive divisors of mn. Hence $F(m)F(n) = F(mn)$, completing the proof. ∎

In order to prove the converse of Theorem 5.10, we must take the equation $F(n) = \sum_{d|n} f(d)$ and derive an equation for f. This will be done by a device known as the *Möbius Inversion Formula*, and it requires the use of a new arithmetic function μ (the Greek letter mu).

Definition. Let $n > 1$ and let $n = p_1^{e_1} p_2^{e_2} \cdots p_s^{e_s}$ be the prime decomposition of n. Define $\mu(n)$ in the following way: If $e_1 = e_2 = \cdots = e_s = 1$, then let $\mu(n) = (-1)^s$. If, however, any of $e_1, e_2, \ldots, e_s$ are greater than 1, then let $\mu(n) = 0$. Finally, let $\mu(1) = 1$.

The definition of μ is not particularly elegant, but it is just what Möbius needed to make his formula work. For practice, $\mu(6) = (-1)^2 = 1$ since $6 = 2^1 \cdot 3^1$; $\mu(12) = 0$ since $12 = 2^2 \cdot 3^1$; $\mu(42) = (-1)^3$ since $42 = 2^1 \cdot 3^1 \cdot 7^1$; and $\mu(250) = 0$ since $250 = 2^1 \cdot 5^3$.

The next two lemmas tell us what we need to know about μ.

Lemma 5.11. The function μ is a multiplicative function.

Proof. Let m and n be positive integers with $(m, n) = 1$. If $m = 1$, then $\mu(mn) = \mu(n) = 1\mu(n) = \mu(1)\mu(n) = \mu(m)\mu(n)$. Similarly, if $n = 1$, $\mu(m)\mu(n) = \mu(mn)$. Now suppose that $m > 1$ and $n > 1$ and let the prime decompositions of m and n be $m = p_1^{e_1} \cdots p_s^{e_s}$, $n = q_1^{d_1} \cdots q_t^{d_t}$. Since $(m, n) = 1$, $p_i \ne q_j$ for $1 \le i \le s$, $1 \le j \le t$, and so the prime decomposition of mn is (except for order)

$$mn = p_1^{e_1} \cdots p_s^{e_s} q_1^{d_1} \cdots q_t^{d_t}$$

Now if some e_i is greater than 1, then both $\mu(m) = 0$ and $\mu(mn) = 0$, so that $\mu(mn) = 0 = 0 \cdot \mu(n) = \mu(m)\mu(n)$. Similarly, if some d_j is greater than 1, then

$\mu(mn) = \mu(m)\mu(n)$. Thus suppose that $e_1 = \cdots = e_s = 1 = d_1 = \cdots = d_t$. Then $\mu(m) = (-1)^s$, $\mu(n) = (-1)^t$, and $\mu(mn) = (-1)^{s+t}$ (since there are $s + t$ distinct primes in the decomposition of mn). Then $\mu(mn) = (-1)^{s+t} = (-1)^s \cdot (-1)^t = \mu(m)\mu(n)$. ∎

Lemma 5.12. If $n \geq 1$, then $\sum_{d|n}\mu(d) = 0$.

Proof. Let G be the arithmetic function defined by $G(n) = \sum_{d|n}\mu(d)$. By Lemma 5.11 and Theorem 5.10, G is multiplicative. Thus if $n = p_1^{e_1}\cdots p_s^{e_s}$ is the prime decomposition of n, then $G(n) = G(p_1^{e_1})G(p_2^{e_2})\cdots G(p_s^{e_s})$. In order to show that $G(n) = 0$, as desired, it is obviously enough to show that $G(p^e) = 0$ for p prime and $e \geq 1$. Now the positive divisors of p^e are $1, p, p^2, \ldots, p^e$. Therefore, $G(p^e) = \sum_{d|p^e}\mu(d) = \mu(1) + \mu(p) + \mu(p^2) + \cdots + \mu(p^e) = 1 + (-1)^1 + 0 + \cdots + 0 = 0$ and we are done. ∎

Theorem 5.13 (Möbius Inversion Formula). Let f be an arithmetic function, and for $n \geq 1$, let $F(n) = \sum_{d|n} f(d)$. Then $f(n) = \sum_{d|n}\mu(d)F(n/d)$.

EXAMPLE. It is instructive to first see how 5.13 works with $n = 12$. We consider Table 5-1, whose rows and columns are both labeled by the positive divisors of 12.

In Table 5-1 the space in the d-row, c-column is left empty if $dc \nmid 12$, and contains $\mu(d)f(c)$ if $dc \mid 12$. For example, the space in the 3-row, 3-column is blank because $3 \cdot 3 \nmid 12$.

We will add together all the entries in this table, by summing each row and then totaling these row sums. For example, if we sum the 2-row, we get

$$\mu(2)f(1) + \mu(2)f(2) + \mu(2)f(3) + \mu(2)f(6) = \mu(2)[f(1) + f(2) + f(3) + f(6)]$$

Now 1, 2, 3, and 6 are all the positive divisors of 6, so by definition $f(1) + f(2) + f(3) + f(6) = F(6)$. Thus the sum of the 2-row is just $\mu(2)F(6) = \mu(2)F(12/2)$.

Table 5-1 Table for $F(12)$

	1	2	3	4	6	12
1	$\mu(1)f(1)$	$\mu(1)f(2)$	$\mu(1)f(3)$	$\mu(1)f(4)$	$\mu(1)f(6)$	$\mu(1)f(12)$
2	$\mu(2)f(1)$	$\mu(2)f(2)$	$\mu(2)f(3)$		$\mu(2)f(6)$	
3	$\mu(3)f(1)$	$\mu(3)f(2)$		$\mu(3)f(4)$		
4	$\mu(4)f(1)$		$\mu(4)f(3)$			
6	$\mu(6)f(1)$	$\mu(6)f(2)$				
12	$\mu(12)f(1)$					

Similarly the sum of the elements in the d-row is quickly seen to be $\mu(d)F(12/d)$. Now totaling these row sums as d ranges over 1, 2, 3, 4, 6, 12, we get $\mu(1)F(12/1) + \mu(2)F(12/2) + \mu(3)F(12/3) + \mu(4)F(12/4) + \mu(6)F(12/6) + \mu(12)F(12/12) = \sum_{d|12}\mu(d)F(12/d)$. This, then, is the sum of all the elements in our table.

We now find the sum of all the elements in our table in a different way, this time summing each column and then totaling the column sums. For instance, the 3-column sums to

$$\mu(1)f(3) + \mu(2)f(3) + \mu(4)f(3) = [\mu(1) + \mu(2) + \mu(4)]f(3)$$

Now 1, 2, and 4 are all the positive divisors of $4 = 12/3$, so this sum can be written as $[\sum_{c|(12/3)}\mu(c)]f(3)$. Since $12/3 = 4 > 1$, by Lemma 5.12 we have $\sum_{c|(12/3)}\mu(c) = 0$, so that the 3-column sums to 0. Similarly, for $c = 1, 2, 3, 4, 6$ the c-column sums to 0 (using Lemma 5.12 and the fact that $12/c > 1$). This leaves only the 12-column, which contains the single entry $\mu(1)f(12) = 1f(12) = f(12)$. Thus we see that the sum of all the entries in our table is just $f(12)$. Comparing this to our earlier calculation, we conclude that $f(12) = \sum_{d|12}\mu(d)F(12/d)$. This verifies the Möbius Inversion Formula for the special case $n = 12$ (regardless of what function f is being considered).

We now begin the complete proof of the Möbius Inversion Formula. The reader should see that the use of the double sigma notation is nothing other than an abbreviated method of repeating the same argument that we have just given.

Proof (of Theorem 5.13). Consider the sum $\sum_{dc|n}\mu(d)f(c)$. [That is, the sum of all terms $\mu(d)f(c)$ as d and c take on all possible positive values satisfying $dc\,|\,n$.] We compute this sum in two ways. First, for a fixed positive d with $d\,|\,n$, it is obvious that $dc\,|\,n$ if and only if $c\,|\,(n/d)$ (Exercise 2.A.22). Consequently,

$$\sum_{dc|n}\mu(d)f(c) = \sum_{d|n}\mu(d)\left[\sum_{c|(n/d)} f(c)\right] = \sum_{d|n}\mu(d)F(n/d)$$

because by assumption $F(n/d) = \sum_{c|(n/d)} f(c)$. (*Note:* This much of the proof corresponds to summing the rows and then totaling the row sums in our example. The argument in the next paragraph corresponds to summing the columns and then totaling the column sums.)

For a fixed positive c with $c\,|\,n$, we have $dc\,|\,n$ if and only if $d\,|\,(n/c)$, so $\sum_{dc|n}\mu(d)f(c) = \sum_{c|n}[\sum_{d|(n/c)}\mu(d)]f(c)$. If $c < n$, then $n/c > 1$ and by Lemma 5.12, $\sum_{d|(n/c)}\mu(d) = 0$. Therefore, in this last sum, every term vanishes except the term arising from $c = n$, which is $(\sum_{d|(n/n)}\mu(d))f(n) = (\sum_{d|1}\mu(d))f(n) = \mu(1)f(n) = 1f(n) = f(n)$. That is, $\sum_{dc|n}\mu(d)f(c) = f(n)$. Comparing this to our earlier calculation, $f(n) = \sum_{d|n}\mu(d)F(n/d)$, and the Möbius Inversion Formula is proved. ∎

We now prove the converse of Theorem 5.10.

Theorem 5.14. Let f be an arithmetic function and let F be the arithmetic function defined by $F(n) = \sum_{d|n} f(d)$. If F is multiplicative, then f is multiplicative.

Proof. Let m and n be positive integers with $(m, n) = 1$. We must show that $f(mn) = f(m)f(n)$. Suppose that $a_1, \ldots, a_r$ are the positive divisors of m, and $b_1, \ldots, b_t$ are the positive divisors of n. By the Möbius Inversion Formula, $f(m) = \mu(a_1)F(m/a_1) + \cdots + \mu(a_r)F(m/a_r)$, and $f(n) = \mu(b_1)F(n/b_1) + \cdots + \mu(b_t)F(n/b_t)$. Therefore,

$$\begin{aligned} f(m)f(n) = {} & \mu(a_1)\mu(b_1)F(m/a_1)F(n/b_1) + \cdots + \mu(a_1)\mu(b_t)F(m/a_1)F(n/b_t) \\ & + \mu(a_2)\mu(b_1)F(m/a_2)F(n/b_1) + \cdots + \mu(a_2)\mu(b_t)F(m/a_2)F(n/b_t) \\ & \vdots \\ & + \mu(a_r)\mu(b_1)F(m/a_r)F(n/b_1) + \cdots + \mu(a_r)\mu(b_t)F(m/a_r)F(n/b_t) \end{aligned}$$

Since $a_i|m$ and $b_j|n$ and $(m, n) = 1$, we have $(a_i, b_j) = 1$ for $1 \le i \le r$, $1 \le j \le t$. Also, $(m/a_i)|m$ and $(n/b_j)|n$, so that $(m/a_i, n/b_j) = 1$. Furthermore, μ is multiplicative by Lemma 5.11, and F is multiplicative by assumption. Therefore, the sum equals

$$\begin{aligned} & \mu(a_1b_1)F(mn/a_1b_1) + \cdots + \mu(a_1b_t)F(mn/a_1b_t) \\ & \quad + \mu(a_2b_1)F(mn/a_2b_1) + \cdots + \mu(a_2b_t)F(mn/a_2b_t) \\ & \quad \vdots \\ & \quad + \mu(a_rb_1)F(mn/a_rb_1) + \cdots + \mu(a_rb_t)F(mn/a_rb_t) \end{aligned}$$

However, by Lemma 5.5, $\{a_ib_j | 1 \le i \le r, 1 \le j \le t\}$ is exactly the set of positive divisors of mn. Therefore, this last sum is just $\sum_{d|mn} \mu(d)F(mn/d)$. Finally, by Theorem 5.13, $\sum_{d|mn} \mu(d)F(mn/d) = f(mn)$. We have shown that $f(m)f(n) = f(mn)$, as desired. ∎

We close this chapter with a striking result which is an easy consequence of the Möbius Inversion Formula.

Proposition 5.15. For any integer $n \ge 1$, $\sum_{d|n} \mu(d)\tau(n/d) = 1$.

Proof. Let $f(n) = 1$ for all n. Then $\sum_{d|n} f(d)$ is just the sum of a certain number of 1's, that number being the number of positive divisors of n. That is, $\sum_{d|n} f(n) = \tau(n)$. By the Möbius Inversion Formula, $\sum_{d|n} \mu(d)\tau(n/d) = f(n) = 1$. ∎

EXERCISES 5.A

1. Calculate: $\varphi(7)$, $\varphi(55)$, $\varphi(108)$.
2. Calculate: $\sigma(7)$, $\sigma(55)$, $\sigma(108)$.
3. Calculate: $\tau(7)$, $\tau(55)$, $\tau(108)$.
4. Verify that $\varphi(nm) = \varphi(n)\varphi(m)$ for the following pairs of relatively prime integers: $n = 5, m = 7$; $n = 5, m = 14$; $n = 15, m = 14$.
5. Proceed as in 5.A.4, replacing φ by σ.
6. Proceed as in 5.A.4, replacing φ by τ.
7. Give examples to show that $\varphi(mn) \neq \varphi(m)\varphi(n)$, $\sigma(mn) \neq \sigma(m)\sigma(n)$, and $\tau(mn) \neq \tau(m)\tau(n)$.
8. Evaluate the following expressions: $\sum_{d|15} \varphi(d)$, $\tau(\varphi(15))$, $\varphi(\tau(15))$.
9. Verify that for $p = 2, 3, 5, 7$, and 13, $2^p - 1$ is a Mersenne prime while $2^{11} - 1$ is not.
10. Verify that 641 divides $2^{32} + 1 = 4{,}294{,}967{,}297$. (*Note:* $32 = 2^5$.)
11. Call a function f, defined on positive integers, an *additive function* if $f(a + b) = f(a) + f(b)$ for all a and b. Are any of φ, σ, or τ additive?
12. Find all positive integers m such that $\varphi(2m) = 2\varphi(m)$.
13. Show that if $m > 2$, then $\varphi(m)$ is even.
14. Show that if $m = p_1^{e_1}p_2^{e_2}\cdots p_k^{e_k}$, then $\varphi(m) = m(1 - 1/p_1)\cdots(1 - 1/p_k)$.
15. Are there any positive integers such that $\sigma(m) = \tau(m)$? $\sigma(m) = \varphi(m)$?
16. Let σ_1 be the function defined by $\sigma_1(m) = \sum_{d|n} \sigma(n)$. (By Theorem 5.10, σ_1 is multiplicative.) Evaluate: $\sigma_1(6)$, $\sigma_1(20)$, $\sigma_1(24)$. Can you find an expression for $\sigma_1(n)$ in terms of the prime factorization of n?
17. Proceed as in Exercise 5.A.16, replacing σ by τ.
18. Let $m \geq 2$ and $n \geq 2$. Suppose that their prime decompositions are $m = p_1^{e_1}\cdots p_k^{e_k}$ and $n = q_1^{f_1}\cdots q_t^{f_t}$. In the proof of Theorem 5.4, we used the fact that if $(m, n) = 1$, then the primary decomposition of mn is (except for order) $mn = p_1^{e_1}p_2^{e_2}\cdots p_k^{e_k}q_1^{f_1}q_2^{f_2}\cdots q_t^{f_t}$. If $(m, n) \neq 1$, why can we not say this and how does it show that $\varphi(mn) \neq \varphi(m)\varphi(n)$?
19. Verify the truth of Lemma 5.5 for $m = 12$, $n = 35$.
20. Suppose that $\varphi_1(n) = \sum_{d|n} \varphi(d)$. Compute $\varphi_1(1)$, $\varphi_1(2)$, $\varphi_1(3)$, $\varphi_1(6)$, $\varphi_1(9)$, and $\varphi_1(18)$.
21. Suppose that $\varphi_2(n) = \sum_{d|n} \varphi_1(n)$ (with φ_1 as in Exercise 5.A.20). Compute $\varphi_2(3)$, $\varphi_2(6)$, and $\varphi_2(9)$.
22. Verify that $\mu(mn) = \mu(m)\mu(n)$ for each of the following relatively prime pairs of integers: $m = 6, n = 35$; $m = 5, n = 21$; and $m = 14, n = 25$.
23. Verify Lemma 5.12 for $n = 2$, $n = 4$, $n = 6$, and $n = 24$.
24. Following the example for $n = 12$ given in the text, give a detailed proof of Theorem 5.13 for $n = 24$.

25. By calculating both sides of the equation, verify that the Möbius Inversion Formula holds for $f = \varphi$, $F = \varphi_1$ (as defined in Exercise 5.A.20), and $n = 18$. That is, verify that $\varphi(18) = \sum_{d|18} \mu(d)\varphi_1(18/d)$.

26. Verify Proposition 5.15 for $n = 6$, $n = 12$, $n = 18$.

27. Write out the proof of Theorem 5.13 as it would appear if you did not use the sigma notation.

28. Write the proof of Theorem 5.14 using sigma notation.

EXERCISES 5.B

1. Give a different proof of Lemma 5.2 by writing the numbers $1, 2, \ldots, p^e$ in an array with p columns and p^{e-1} rows.

2. Show that if $a \mid b$, then $\varphi(ab) = a\varphi(b)$. Is the converse true?

3. Prove: If $2^k - 1$ is a prime, then k is a prime (*Hint:* If $k = ab$, then $2^a - 1$ divides $2^{ab} - 1$.)

4. Show that if n is an even perfect number, then $8n + 1$ is a square.

5. Show that if n is an even perfect number greater than 6, then $n \equiv 1 \bmod 9$. (*Hint:* If $a \geq 1$, then a^2 is congruent to 0, 1, 4, or 7 modulo 9; use the previous exercise.)

6. Let f be the function defined by $f(d) = d$ for all positive integers d. Show that f is a multiplicative function and use this fact to deduce Theorem 5.6 from Theorem 5.10.

7. Deduce Theorem 5.4 from Theorem 5.10.

8. Show that if p is a prime, then for any integer $e \geq 1$,

$$\sum_{d|p^e} \varphi(d) = p^e$$

9. Use Exercise 5.B.8 and Theorem 5.10 to show that $\sum_{d|n} \varphi(d) = n$ for all positive integers n.

10. Let f and g be multiplicative functions such that $f(p^k) = g(p^k)$ for all primes p and all integers $k \geq 1$. Show that $f(n) = g(n)$ for all integers $n \geq 1$.

11. Call a function f *strongly multiplicative* if $f(mn) = f(n)f(m)$ for all positive integers n and m. Find a function f which is strongly multiplicative but such that $F(n) = \sum_{d|n} f(n)$ is not strongly multiplicative (*Hint:* What is the easiest strongly multiplicative function you know?)

12. Determine whether or not any of the following functions are strongly multiplicative:

(a) $P(n) = \sum_{d|n} \varphi(d)$ (b) $S(n) = \sum_{d|n} \sigma(d)$

(c) $T(n) = \sum_{d|n} \tau(d)$

13. Let f and g be multiplicative functions and let fg be the function defined by $(fg)(n) = f(n)g(n)$ for all positive integers n.

(a) Show that fg is multiplicative.

(b) Can you express $H(n) = \sum_{d|n}(fg)(d)$ in terms of $F(n) = \sum_{d|n} f(d)$ and $G(n) = \sum_{d|n} g(d)$?

14. An integer n is *triangular* if $n = 1 + 2 + \cdots + k$ for some integer $k \geq 1$ (the first four triangular numbers are 1, 3, 6, and 10). Show that every even perfect number is triangular.

15. Show that $n \geq 1$ is triangular if and only if $8n + 1$ is a square. (See also Exercises 5.B.14 and 4.)

16. Prove that $\sum_{d|n} \mu(d)\sigma(n/d) = n$.

6

Solving Congruences

THIS chapter is divided into four sections, each of which deals with solving congruences. In Section 6.1 we solve the congruence $aX \equiv c \bmod m$, where a, c, and m are given integers, $m > 0$. Section 6.2 treats the problem of finding a simultaneous solution to the system of congruences

$$a_i X \equiv c_i \bmod m_i, \qquad 1 \leq i \leq k$$

In Section 6.3 we attack the congruence $X^2 \equiv a \bmod m$. Section 6.4 gives an important result concerning the number of solutions of the congruence

$$h(X) \equiv 0 \bmod p$$

where $h(X)$ is any polynomial and p is a prime.

In one sense each of these problems under consideration is easier to handle than the corresponding problem with equality. For example, in the case of the equation $3X = 4$, we are seeking an integer X such that 3 times X equals 4; of course, there is none. However, in the case of the congruence $3X \equiv 4 \bmod 7$, we only seek an integer X such that 3 times X minus 4 be divisible by 7. There are many candidates, such as $X = -1$, $X = 6$, or $X = 27$. Hence the problem is, in many cases, not one of finding solutions, but rather one of systematically organizing the solutions.

We have devoted much effort to the section where we are concerned with solving the congruence $X^2 \equiv a \bmod m$. We show that it is enough to be able to solve congruences of the form $X^2 \equiv a \bmod p$, *where p is a prime.* Our effort is intended to not only solve $X^2 \equiv a \bmod m$ but also to convince you that the problem of solving $X^2 \equiv a \bmod p$ with p prime is an important one. To illustrate how truly important this problem is would require excursions into some of the loftiest of abstract mathematics and is well beyond the intent and scope of this book. We do not solve the congruence $X^2 \equiv a \bmod p$ in this chapter; Chapter 9 is devoted to the consideration of this congruence and one of the most noteworthy and beautiful theorems of all mathematics. But let's begin at the beginning. . . .

6.1 The Congruence $aX \equiv b \bmod m$

We cannot overstress the value of looking at examples. Let us look at $6X \equiv 3 \bmod 15$. Examining $x = 0, 1, 2, \ldots, 14$, we find that 3, 8, and 13 are solutions. Now since we are working modulo 15, obviously 18 is also a solution, since $18 \equiv 3 \bmod 15$, while 17 is not a solution, since $17 \equiv 2 \bmod 15$ and 2 is not a solution. Thus it is easy to see that the entire list of solutions reads,

$$\ldots, -12, -7, -2, 3, 8, 13, 18, 23, \ldots$$

Notice that the solutions are separated by 5. In other words, the solutions are exactly the set $\{x \mid x \equiv 3 \bmod 5\}$. Also, three of the solutions are between 0 and $m - 1 = 14$.

Now let us consider $21X \equiv 14 \bmod 28$. Examining $x = 0, 1, 2, \ldots, 27$, we find that 2, 6, 10, 14, 18, 22, and 26 are solutions. As we are working modulo 28, the entire list of solutions must be

$$\ldots, -10, -6, -2, 2, 6, 10, 14, 18, 22, 26, 30, 34, \ldots$$

Thus the solutions are $\{x \mid x \equiv 2 \bmod 4\}$, and seven of them fall between 0 and $m - 1 = 27$.

By now you have probably guessed that for $aX \equiv c \bmod m$, if you let $d = (a, m)$, then there are d solutions between 0 and $m - 1$, and all solutions are spaced a distance m/d apart. You are *almost* right.

Consider $8X \equiv 3 \bmod 12$. Trying $x = 0, 1, 2, \ldots, 11$, we find that this congruence does not have any solutions! You have seen this sort of thing happen before, and now you probably want to improve your guess by stipulating that it works when $d \mid c$, and adding that if $d \nmid c$, then there are no solutions. Checking more examples will give you confidence that your improved guess is correct.

We now prove what you have observed to be true. To do so we will make use of our knowledge of Diophantine equations (Chapter 2). As we already mentioned in Chapter 4, if x is a solution of $aX \equiv c \bmod m$, then $m \mid ax - c$ and so $ax - c = my$ for some $y \in \mathbb{Z}$. Obviously, this means that $x, -y$ is a solution of the Diophantine equation $aX + mY = c$. Let us recall what we know about such equations.

Theorem 6.1 (Theorem 2.14 revisited). Let $a, b, c \in \mathbb{Z}$ and $d = (a, b)$.

(i) If $d \nmid c$, then $aX + bY = c$ has no solutions.

(ii) If $d \mid c$, then $aX + bY = c$ has infinitely many solutions, and they are given by $X = x_0 + (b/d)t$, $Y = y_0 - (a/d)t$, where x_0, y_0 is a solution and t is any integer.

We translate this into the language of congruences to learn about $aX \equiv c \bmod m$.

Theorem 6.2. Let $m > 0$, $a \in \mathbb{Z}$, and $d = (a, m)$.

(i) If $d \nmid c$, then $aX \equiv c \bmod m$ has no solutions.

(ii) If $d \mid c$, then $aX \equiv c \bmod m$ has solutions. If x_0 is one solution, then the set of all solutions is exactly $\{x \mid x \equiv x_0 \bmod m/d\}$.

(iii) If $d \mid c$, then exactly d of the solutions of $aX \equiv c \bmod m$ fall between 0 and $m - 1$; and these give all the distinct solutions modulo m.

Proof. As already shown, x is a solution of $aX \equiv c \bmod m$ if and only if there is $-y \in \mathbb{Z}$ with $x, -y$ a solution of $aX + mY = c$. If $d \nmid c$, then Theorem 6.1 tells us that there are no such pairs $x, -y$. Thus $aX \equiv c \bmod m$ has no solutions and (i) is proved.

If $d \mid c$, then Theorem 6.1 tells us that there are solutions to $aX + mY = c$ and they are $x = x_0 + (m/d)t$, $y = y_0 - (a/d)t$ with x_0, y_0 being a solution and $t \in \mathbb{Z}$. Therefore, $\{x_0 + (m/d)t \mid t \in \mathbb{Z}\}$ is the set of solutions of $aX \equiv c \bmod m$. However $x = x_0 + (m/d)t$ if and only if $x \equiv x_0 \bmod m/d$, so the solutions of $aX \equiv c \bmod m$ are just $\{x \mid x \equiv x_0 \bmod m/d\}$. This proves (ii).

Finally for (iii), if $d \mid c$, we know that $aX \equiv c \bmod m$ has solutions and so let x_0 be the smallest solution satisfying $x_0 \geq 0$. By (ii), $x_0 - m/d$ is also a solution, so by our choice of x_0 we must have $x_0 - m/d < 0$. That is, $x_0 < m/d$. We now have $0 \leq x_0 < m/d$. We now leave it to the reader to argue that

$$x_0, x_0 + m/d, x_0 + 2(m/d), \ldots, x_0 + (d - 1)(m/d)$$

are all the solutions modulo m, and thus complete the proof. ∎

As you might expect, the solutions of $aX = c \bmod m$ that fall between 0 and $m - 1$ are called the *solutions modulo m of* $aX \equiv c \bmod m$.

Theorem 6.2 shows us how to find all solutions of $aX \equiv c \bmod m$ if we have any one solution x_0 in hand. How do we find x_0? Well, since for some $y_0 \in Z$, $x_0, -y_0$ will be a solution of $aX + mY = c$, any of the methods of Chapter 2 for finding a solution to this equation can be used. Let us illustrate two other methods.

EXAMPLES.

(a) Consider the congruence

$$18X \equiv 30 \bmod 66$$

Since $(18, 66) = 6$ and $6 \mid 30$, there are solutions. We divide the given congruence by 6 and obtain

$$3X \equiv 5 \bmod 11$$

(We trust that the reader can prove that the new congruence will have the same solutions as the old one.) Now $(3, 11) = 1$, so 3 has a multiplicative inverse modulo 11. Using one of the methods from Chapter 4, or inspection, we find that $3 \cdot 4 \equiv 1 \bmod 11$. Multiplying $3X \equiv 5 \bmod 11$ by 4 yields $1 \cdot X \equiv 4 \cdot 5 \equiv 9 \bmod 11$. Thus $x_0 = 9$ is a solution to our original congruence $18X \equiv 30 \bmod 66$.

(b) Another method of finding a solution of $18X \equiv 30 \bmod 66$ is to divide through by $6 = (18, 30)$. This yields $3X \equiv 5 \bmod 11$. Now add multiples of 11 to 5 until you reach a multiple of 3. Thus

$$3X \equiv 5 \equiv 16 \equiv 27 \bmod 11$$

Since $27 = 3 \cdot 9$, take $x_0 = 9$.

6.2 Systems of Linear Congruences

Suppose that we have congruences

$$6X \equiv 4 \bmod 8$$
$$3X \equiv 2 \bmod 5$$

The methods of the previous section tell us that we can solve each of these separately and the question we ask is whether we can find one x that will solve both. Such an x will be called a *simultaneous solution of the system of congruences.* You can quickly verify that $\{x \mid x \equiv 2 \bmod 4\}$ describes all solutions of the first congruence, and $\{x \mid x \equiv 4 \bmod 5\}$ describes all the solutions to the second congruence. Thus we seek an integer that is congruent to 2 modulo 4 and congruent to 4 modulo 5. An example of such an integer is $x = 14$; thus there is a simultaneous solution to the system.

In this example we were able to find a simultaneous solution to both congruences. It would be quite useful to know when solutions exist before embarking on what could possibly be a frustrating search. Thus we would like to develop criteria, such as in Theorem 6.2, which tell us whether or not a system of linear congruences has a simultaneous solution. Here is one such, which is an obvious consequence of Theorem 6.2.

Proposition 6.3. Consider the system of congruences

$$\begin{aligned} a_1X &\equiv c_1 \bmod m_1 \\ a_2X &\equiv c_2 \bmod m_2 \\ &\vdots \\ a_kX &\equiv c_k \bmod m_k \end{aligned}$$

and let $d_i = (a_i, m_i)$ for $i = 1, 2, \ldots, k$. If for some $j \in \mathbb{Z}$, $1 \leq j \leq k$, $d_j \nmid c_i$, then no simultaneous solutions exist.

The moral of this proposition is that if you cannot have solutions for one congruence, you certainly should not expect to be able to solve a system that includes this congruence. Thus a first step in solving any system of congruences is to first see whether or not each individual congruence is solvable. This in itself will not guarantee existence of simultaneous solutions. For example, both of the congruences

$$6X \equiv 4 \bmod 8$$
$$9X \equiv 3 \bmod 12$$

are solvable. The first has solution set $\{x \mid x \equiv 2 \bmod 4\}$ while the second has solution set $\{x \mid x \equiv 3 \bmod 4\}$, as you can easily verify. It is clear that no x can satisfy both congruences, since there are no integers that are congruent to both 2 and 3 modulo 4. Thus while systems may trivially fail to have solutions because they do not make Proposition 6.3 happy, they may also fail for deeper reasons.

In searching for criteria and methods of solutions, we first have a reduction technique.

Theorem 6.4. For the system of congruences

$$\begin{gathered} a_1X \equiv c_1 \bmod m_1 \\ \vdots \\ a_kX \equiv c_k \bmod m_k \end{gathered}$$

let $d_i = (a_i, m_i)$ for $1 \leq i \leq k$, and suppose that x_i is a solution of the ith congruence, for $1 \leq i \leq k$. Then x is a simultaneous solution of the system if and only if x is a solution of the following system:

$$\begin{gathered} X \equiv x_1 \bmod \left(\frac{m_1}{d_1}\right) \\ \vdots \\ X \equiv x_k \bmod \left(\frac{m_k}{d_k}\right) \end{gathered}$$

Proof. Since x_1 is a solution of $a_1X \equiv c_1 \bmod m_1$ and $d_1 = (a_1, m_1)$, Theorem 6.2 tells us that the set of all solutions of $a_1X \equiv c_1 \bmod m_1$ is $\left\{x \mid x \equiv x_1 \bmod \left(\frac{m_1}{d_1}\right)\right\}$. Similarly for any $i \in \mathbb{Z}$, $1 \leq i \leq k$, the set of all solutions of the ith congruence $a_iX \equiv c_i \bmod m_i$ is the set $\left\{x \mid x \equiv x_i \bmod \left(\frac{m_i}{d_i}\right)\right\}$.

Thus if x is a simultaneous solution of all the equations $a_i X \equiv c_i \bmod m_i$ for $1 \leq i \leq k$, then x must belong to each of the sets $\left\{x \middle| x \equiv x_i \bmod \left(\frac{m_i}{d_i}\right)\right\}$ for $1 \leq i \leq k$. Put another way, x is a solution of the system

$$\begin{gathered} X \equiv x_1 \bmod \left(\frac{m_1}{d_1}\right) \\ \vdots \\ X \equiv x_k \bmod \left(\frac{m_k}{d_k}\right) \end{gathered}$$

Now for the converse. If x is a simultaneous solution of

$$\begin{gathered} X \equiv x_1 \bmod \left(\frac{m_1}{d_1}\right) \\ \vdots \\ X \equiv x_k \bmod \left(\frac{m_k}{d_k}\right) \end{gathered}$$

then, since $x \equiv x_i \bmod \left(\frac{m_i}{d_i}\right)$ for each $i = 1, \ldots, k$, by Theorem 6.2 we have that x is a solution of each of $a_i X \equiv c_i \bmod m_i$, for $1 \leq i \leq k$. Hence x is a simultaneous solution of the original system. ∎

Applying this result to our earlier example of

$$\begin{gathered} 6X \equiv 4 \bmod 8 \\ 9X \equiv 3 \bmod 12 \end{gathered}$$

we have $\frac{m_1}{d_1} = \frac{8}{2} = 4$ and $\frac{m_2}{d_2} = \frac{12}{3} = 4$. Theorem 6.4 tells us that the system

$$\text{(A)} \qquad \begin{gathered} 6X \equiv 4 \bmod 8 \\ 9X \equiv 3 \bmod 12 \end{gathered}$$

has the same set of solutions as the system

$$\text{(B)} \qquad \begin{gathered} X \equiv 2 \bmod 4 \\ X \equiv 3 \bmod 4 \end{gathered}$$

because $\{x | x \equiv 2 \bmod 4\}$ is the set of all solutions to $6X \equiv 4 \bmod 8$, while $\{x | x \equiv 3 \bmod 4\}$ is the set of all solutions to $9X \equiv 3 \bmod 12$. Clearly, the system (B) has no solutions and so the system (A) has no solutions.

In the case of our first example, we had the original system

$$\text{(A}'\text{)} \qquad \begin{aligned} 6X &\equiv 4 \bmod 8 \\ 3X &\equiv 2 \bmod 5 \end{aligned}$$

and the system whose solution set was identical with that of (A′) was the system

$$\text{(B}'\text{)} \qquad \begin{aligned} X &\equiv 2 \bmod 4 \\ X &\equiv 4 \bmod 5 \end{aligned}$$

The effect of Theorem 6.4 is to reduce the problem from a system where arbitrary coefficients a_i of X are allowed, to a system where the only coefficient of X is 1. Thus we can turn our attention now to systems of the form

$$(*) \qquad \begin{aligned} X &\equiv c_1 \bmod m_1 \\ &\vdots \\ X &\equiv c_k \bmod m_k \end{aligned}$$

We will first look at a special type of system which always has a solution. We consider a system

$$\begin{aligned} X &\equiv c_1 \bmod m_1 \\ X &\equiv c_2 \bmod m_2 \end{aligned}$$

with the special assumption that $(m_1, m_2) = 1$.

Example Consider the two congruences

$$\begin{aligned} X &= 7 \bmod 11 \\ X &\equiv 10 \bmod 18 \end{aligned}$$

Noticing that $(11, 18) = 1$, we reason as follows: If we are to have $x \equiv 7 \bmod 11$, then

$$x = 11t_1 + 7$$

for some $t_1 \in \mathbb{Z}$. If also $x \equiv 10 \bmod 18$, then

$$11t_1 + 7 \equiv 10 \bmod 18$$

Thus

$$11t_1 \equiv 3 \bmod 18$$

Since $(11, 18) = 1$, the previous section assures us that there are such integers t_1, and give us ways of finding them. For instance, since the inverse of 11 modulo 18 is 5, that is, since $5 \cdot 11 \equiv 1 \bmod 18$, we see that

$$t_1 \equiv (5 \cdot 11)t_1 \equiv 5 \cdot 3 = 15 \bmod 18$$

Hence all such t_1 are given by the expression

$$t_1 = 18t_2 + 15, \qquad \text{where } t_2 \in \mathbb{Z}$$

Since $x = 11t_1 + 7$, we have

$$x = 11(18t_2 + 15) + 7 = (11 \cdot 18)t_2 + 172$$

Therefore, as t_2 varies through $\mathbb{Z}$, this gives all solutions to our pair of congruences. Of course, we could also express these solutions as

$$x \equiv 172 \bmod (11 \cdot 18)$$

Now suppose that we have a system of three congruences such as

$$\text{(C)} \qquad \begin{aligned} X &\equiv 7 \bmod 11 \\ X &\equiv 10 \bmod 18 \\ X &\equiv 3 \bmod 7 \end{aligned}$$

in which we have $(11, 18) = 1$, $(11, 7) = 1$, and $(18, 7) = 1$. We have just seen that the first pair of congruences in this system has solution

$$x \equiv 172 \bmod (11 \cdot 18)$$

Therefore, it is obvious that the system (C) has exactly the same solutions as

$$\text{(D)} \qquad \begin{aligned} X &\equiv 172 \bmod (11 \cdot 18) \\ X &\equiv 3 \bmod 7 \end{aligned}$$

Notice that $(11 \cdot 18, 7) = 1$ (see Exercise 2.B.9). To solve the system (D) we reason that if $x \equiv 172 \bmod (11 \cdot 18)$, then

$$x = (11 \cdot 18)t_2 + 172, \qquad \text{where } t_2 \in \mathbb{Z}$$

Since we also want $x \equiv 3 \bmod 7$, we want

$$(11 \cdot 18)t_2 + 172 \equiv 3 \bmod 7$$

Thus

$$(11 \cdot 18)t_2 \equiv -169 \equiv 6 \bmod 7$$

Now since $11 \cdot 18 = 198 \equiv 2 \bmod 7$, and since $4 \cdot 2 \equiv 1 \bmod 7$, we have

$$t_2 \equiv 4(11 \cdot 18)t_2 \equiv 4 \cdot 6 = 24 \equiv 3 \bmod 7$$

That is,

$$t_2 = 7t_3 + 3$$

Since $x = (11 \cdot 18)t_2 + 172$,

$$x = 11 \cdot 18(7t_3 + 3) + 172 = 11 \cdot 18 \cdot 7t_3 + 766$$

That is, the solutions to our system (D), and hence also our system (C), are described by the congruence

$$x \equiv 766 \bmod 11 \cdot 18 \cdot 7$$

Were there a fourth congruence in our system, we would repeat the process another time. We formalize this using the following definition.

Definition. We say that the integers $m_1, m_2, \ldots, m_k$ are *pairwise relatively prime* if for each pair of distinct subscripts i, j we have $(m_i, m_j) = 1$.

For example, 11, 18, and 7 are pairwise relatively prime, but 15, 7, and 18 are not since $(15, 18) = 3$.

Theorem 6.5 (The Chinese Remainder Theorem). Suppose that the positive integers $m_1, m_2, \ldots, m_k$ are pairwise relatively prime. Then for any integers $c_1, c_2, \ldots, c_k$, the system of congruences

$$\begin{aligned} X_1 &\equiv c_1 \bmod m_1 \\ X_2 &\equiv c_2 \bmod m_2 \\ &\vdots \\ X_k &\equiv c_k \bmod m_k \end{aligned}$$

has a solution. Furthermore, if x_0 is one solution, then all solutions are given by

$$x \equiv x_0 \bmod m_1 m_2 \cdots m_k$$

Proof. We induct on k, the number of congruences. If $k = 1$, there is nothing to be done. For $k = 2$, we have

$$\begin{aligned} X_1 &\equiv c_1 \bmod m_1 \\ X_2 &\equiv c_2 \bmod m_2 \end{aligned}$$

We want $x = m_1 t_1 + c_1$ for some $t_1 \in \mathbb{Z}$ and also $x \equiv c_2 \bmod m_2$. That is, we want

$$m_1 t_1 + c_1 \equiv c_2 \bmod m_2$$

so that

$$m_1 t_1 \equiv c_2 - c_1 \bmod m_2$$

Because $(m_1, m_2) = 1$ there is an integer such that $u_1 m_1 \equiv 1 \bmod m_2$ (Theorem 4.3). Thus we want

$$t_1 \equiv (u_1 m_1) t_1 \equiv u_1 (c_2 - c_1) \bmod m_2$$

That is,

$$t_1 = m_2 t_2 + u_1(c_2 - c_1), \qquad t_2 \in \mathbb{Z}$$

Since $x = m_1 t_1 + c_1$, we have

$$x = m_1(m_2 t_2 + u_1(c_2 - c_1)) + c_1 = m_1 m_2 t_2 + m_1 u_1(c_2 - c_1) + c_1$$

That is,

$$x \equiv m_1 u_1(c_2 - c_1) + c_1 \bmod m_1 m_2$$

Thus $x_0 = m_1 u_1(c_2 - c_1) + c_1$ is a specific solution of our pair of equations, and $x \equiv x_0 \bmod m_1 m_2$ gives all solutions. This completes the argument for $k = 2$.

Now suppose that our theorem is true for any system containing $k - 1$ or fewer congruences. To treat the system

$$\text{(E)} \qquad \begin{aligned} X &\equiv c_1 \bmod m_1 \\ X &\equiv c_2 \bmod m_2 \\ &\vdots \\ X &\equiv c_k \bmod m_k \end{aligned}$$

We let x' be a solution of

$$\begin{aligned} X &\equiv c_1 \bmod m_1 \\ X &\equiv c_2 \bmod m_2 \end{aligned}$$

By the case $k = 2$ we know that there is such an x' and furthermore that all solutions to this pair of congruences are given by the expression

$$x \equiv x' \bmod m_1 m_2$$

Therefore, our system (E) obviously has the same solutions as

$$\text{(F)} \qquad \begin{aligned} X &\equiv x' \bmod m_1 m_2 \\ X &\equiv c_3 \bmod m_3 \\ &\vdots \\ X &\equiv c_k \bmod m_k \end{aligned}$$

Now by Exercise 3.A.23 it is easy to see that $(m_1 m_2), m_3, \ldots, m_k$ are pairwise relatively prime. Furthermore, the system (F) contains $k - 1$ congruences. Then by induction we have that (F) has a solution x_0 and that all solutions to (F) are given by the congruence

$$x \equiv x_0 \bmod (m_1 m_2) m_3 \ldots m_k$$

However, since (F) and (E) have the same solutions, x_0 is a solution of (E) and all solutions of (E) are given by the expression

$$x \equiv x_0 \bmod m_1 m_2 \cdots m_k$$

This completes the proof of the Chinese Remainder Theorem. ∎

The Chinese Remainder Theorem is a venerable result known to the Chinese in the first century A.D. Its usefulness is not restricted to the integers, for it has important generalizations throughout that modern-day branch of mathematics known as algebra. The proof we have just given works for the integers, but does not work in more abstract situations. Therefore, we would like to indicate another proof which does carry over to other situations. We will not give the entire argument but content ourselves with showing that if $(m_1, m_2) = 1$, then we can solve

$$X \equiv c_1 \bmod m_1$$
$$X \equiv c_2 \bmod m_2$$

Since $(m_1, m_2) = 1$, there are integers u and v with $um_1 + vm_2 = 1$ (Theorem 2.8). We claim that $x_0 = um_1(c_2 - c_1) + c_1$ solves our pair of congruences. Since $m_1 \mid um_1(c_2 - c_1)$, we obviously have $x_0 \equiv c_1 \bmod m_1$. Thus we must show that $x_0 \equiv c_2 \bmod m_2$. Now, using the fact that $um_1 + vm_2 = 1$, we have

$$\begin{aligned} x_0 - c_2 &= um_1(c_2 - c_1) + c_1 - c_2 \\ &= um_1(c_2 - c_1) - (c_2 - c_1) = (um_1 - 1)(c_2 - c_1) \\ &= -vm_2(c_2 - c_1) \end{aligned}$$

Since $x_0 - c_2 = -vm_2(c_2 - c_1)$, we arrive at $x_0 \equiv c_2 \bmod m_2$, as desired

Notice that this clever argument required suspecting that $x_0 = um_1 \cdot (c_2 - c_1) + c_1$ beforehand. The first proof given showed how to arrive at that fact.

Having settled the matter of systems whose moduli are pairwise relatively prime, we now turn to an arbitrary system of the form

$$\begin{gathered} X \equiv c_1 \bmod m_1 \\ \vdots \\ X \equiv c_k \bmod m_k \end{gathered}$$

with no assumption on the moduli. The ideas we will use are similar to those we have just seen.

We consider an example which will motivate our next definition. Suppose that we have the system

$$X \equiv 2 \bmod 4$$
$$X \equiv 3 \bmod 6$$

Now $(4, 6) = 2$. If x satisfies the first congruence, then we must have $4 \mid x - 2$. Since $2 \mid 4$, we then know that $2 \mid x - 2$. Therefore, any solution of $X \equiv 2 \bmod 4$ must necessarily be a solution of the congruence $X \equiv 2 \bmod 2$. Similarly, if x solves the second congruence, then $2 \mid x - 3$ because $2 \mid 6$ and $6 \mid x - 3$. Thus any solution of the second congruence is a solution of $X \equiv 3 \bmod 2$. This shows that if there is a simultaneous solution to the system

$$X \equiv 2 \bmod 4$$
$$X \equiv 3 \bmod 6$$

then it must be a simultaneous solution of the system

$$X \equiv 2 \bmod 2$$
$$X \equiv 3 \bmod 2$$

Clearly, there is no solution to this last system, so there can be none to the original system.

Definition. Consider the system of congruences

$$X \equiv c_1 \bmod m_1$$
$$\vdots$$
$$X \equiv c_k \bmod m_k$$

and for each pair of subscripts i, j let $e_{ij} = (m_i, m_j)$. We say that the system of congruences is *compatible* if it is true that

$$c_i \equiv c_j \bmod e_{ij} \text{ (for all pairs of subscripts } i, j)$$

Example. The system

$$X \equiv 2 \bmod 4$$
$$X \equiv 3 \bmod 6$$

is not compatible because $(4, 6) = 2$ and $2 \not\equiv 3 \bmod 2$. The system of congruences

$$X \equiv -3 \bmod 10$$
$$X \equiv 1 \bmod 6$$
$$X \equiv 2 \bmod 35$$

is compatible because:

1. $(10, 6) = 2$ and $-3 \equiv 1 \bmod 2$.
2. $(10, 35) = 5$ and $-3 \equiv 2 \bmod 5$.
3. $(6, 35) = 1$ and $1 \equiv 2 \bmod 1$.

Notice that the definition of compatibility only considers two congruences at a time. Thus compatibility of the system of three congruences

$$X \equiv -3 \bmod 10$$
$$X \equiv 1 \bmod 6$$
$$X \equiv 2 \bmod 35$$

is equivalent to compatibility of each of the three systems of two congruences

$$\begin{array}{lll} X \equiv -3 \bmod 10 & X \equiv -3 \bmod 10 & X \equiv 1 \bmod 6 \\ X \equiv 1 \bmod 6 & X \equiv 2 \bmod 35 & X \equiv 2 \bmod 35 \end{array}$$

We will use this observation later.

Theorem 6.6. If the system of congruences

$$\begin{gathered} X \equiv c_1 \bmod m_1 \\ \vdots \\ X \equiv c_k \bmod m_k \end{gathered}$$

has a simultaneous solution, then it is compatible.

Proof. Suppose that we have a solution x to the system. Let i, j be any pair of subscripts and $e_{ij} = (m_i, m_j)$; we must establish that $c_i \equiv c_j \bmod e_{ij}$. Now x is a simultaneous solution, so $x \equiv c_i \bmod m_i$ and $m_i \mid x - c_i$. Since $e_{ij} \mid m_i$, we have $e_{ij} \mid x - c_i$, showing that $x \equiv c_i \bmod e_{ij}$. Similarly, $x \equiv c_j \bmod e_{ij}$; hence $c_i \equiv x \equiv c_j \bmod e_{ij}$, and we are finished. ∎

Compatibility provides a test for solutions, since if a system is not compatible, it cannot have a simultaneous solution. We can see quickly that the system

$$X \equiv 14 \bmod 18$$
$$X \equiv 3 \bmod 7$$
$$X \equiv 10 \bmod 12$$

has no solutions, since $(18, 12) = 6$ and $14 \not\equiv 10 \bmod 6$, so the system is not compatible. Thus we have one immediate use of compatibility. What we will show in a while is that the converse of Theorem 6.6 also holds! It will take some effort to do this, but our energies will be well expended, because along the way we shall learn how to solve any compatible system. Our first step deals with systems having two congruences.

Lemma 6.7. The system

$$X \equiv c_1 \bmod m_1$$
$$X \equiv c_2 \bmod m_2$$

has a simultaneous solution if and only if it is compatible.

Proof. We already know from Theorem 6.6 that if the system has a solution, then it is compatible. Therefore, we suppose that the system is compatible and proceed to produce a solution. If we let $e = e_{12} = (m_1, m_2)$, then, by compatibility, $c_1 \equiv c_2 \bmod e$ and so $e | c_1 - c_2$. We pass to the Diophantine equation

$$m_1 Y + m_2 Z = c_1 - c_2$$

This equation has solutions because $e = (m_1, m_2)$ and $e | c_1 - c_2$. Thus there are integers y_0, z_0 such that $m_1 y_0 + m_2 z_0 = c_1 - c_2$, so $c_1 = m_1 y_0 + m_2 z_0 + c_2$. We let $x = m_2 z_0 + c_2$. Then $x - c_2 = m_2 z_0$, so $x \equiv c_2 \bmod m_2$. But also $c_1 = m_1 y_0 + x$, so $x - c_1 = m_1(-y_0)$, and hence we also have $x \equiv c_1 \bmod m_1$. We have thus obtained a simultaneous solution, namely $x = m_2 z_0 + c_2$, to the system. ∎

A nice feature of Lemma 6.7 is that not only does it tell us that if the equations are compatible, then there is a solution, but its proof even shows us how to find one.

EXAMPLE. Let us apply the proof to the congruence system

$$X \equiv -3 \bmod 10$$
$$X \equiv 1 \bmod 6$$

Note that $(10, 6) = 2$ and $-3 \equiv 1 \bmod 2$, so the system is compatible. Lemma 6.7 instructs us to find a solution of $10Y + 6Z = -3 - 1$, and we already know how to solve such an equation. In this case we can take $y = -1, z = 1$ as a solution. Next we let $x = 6z + c_2 = 6(1) + 1 = 7$, and this will be a solution of the system, as you can easily check out.

Of course there will be other solutions besides 7 to this system. It is, in fact, quite an easy matter to find all of them. For x will be a solution of the system exactly when we have both $x \equiv -3 \equiv 7 \bmod 10$ and $x \equiv 1 \equiv 7 \bmod 6$. Thus the set of all solutions to the system

$$X \equiv -3 \bmod 10$$
$$X \equiv 1 \bmod 6$$

is the same as the set $\{x|x \equiv 7 \bmod 10 \text{ and } x \equiv 7 \bmod 6\} = \{x|\text{both } 10 \text{ and } 6 \text{ divide } x - 7\}$. The least common multiple of 10 and 6 is 30. Therefore, saying that both 10 and 6 divide $x - 7$ is the same as saying that $30|(x - 7)$ (Exercise 3.B.14). We then have that the set of all solutions to the system

$$X \equiv -3 \bmod 10$$
$$X \equiv 1 \bmod 6$$

is $\{x|x \equiv 7 \bmod 30\}$. That is, the system of two congruences

$$X \equiv -3 \bmod 10$$
$$X \equiv 1 \bmod 6$$

has the same solution set as the single congruence

$$X \equiv 7 \bmod 30$$

That this method works in general is the content of our next result.

Lemma 6.8. Consider the congruences

$$X \equiv c_1 \bmod m_1$$
$$X \equiv c_2 \bmod m_2$$

and suppose that x_0 is a solution to this system. Let m be the least common multiple of m_1 and m_2. Then x is a solution of the system if and only if x is a solution of the congruence

$$X \equiv x_0 \bmod m$$

Proof. Since an integer x is a solution of both congruences exactly when $x \equiv c_1 \equiv x_0 \bmod m_1$ and $x \equiv c_2 \equiv x_0 \bmod m_2$, the set of solutions to the system is the same as the set $\{x|\text{both } m_1 \text{ and } m_2 \text{ divide } x - x_0\}$. However, m is the least common multiple of m_1 and m_2, so this last set can be described by $\{x|m \text{ divides } x - x_0\}$. Of course, this is nothing more than the solutions of the congruence

$$X \equiv x_0 \bmod m$$ ∎

Having dealt effectively with two congruences, let us now consider a system with $k \geq 2$ congruences. We look at an example first, the system

$$X \equiv -3 \bmod 10$$
$$X \equiv 1 \bmod 6$$
$$X \equiv 2 \bmod 35$$

We have already seen that this system is compatible. Restricting attention to the first two congruences, we see that this system automatically is compatible because it is part of a larger compatible system. Lemma 6.7 tells us that solutions exist and how to find them. We have already carried out this process and found that the set of solutions to the system

$$X \equiv -3 \bmod 10$$
$$X \equiv 1 \bmod 6$$

is the same as the set of solutions to the congruence

$$X \equiv 7 \bmod 30$$

Thus the problem of solving the original system of three congruences is the same as that of solving the system

$$X \equiv 7 \bmod 30$$
$$X \equiv 2 \bmod 35$$

Therefore, instead of having three congruences in our system, we have reduced the problem to that of solving a system with only two congruences. We turn again to Lemma 6.7. Since $(30, 35) = 5$ and $7 \equiv 2 \bmod 5$, this new system is compatible, solutions exist, and Lemma 6.7 tells us how to find them. Before finishing up this problem, however, let us ask an important question. Is the compatibility of the system

$$X \equiv 7 \bmod 30$$
$$X \equiv 2 \bmod 35$$

a consequence of the original compatibility of

$$X \equiv -3 \bmod 10$$
$$X \equiv 1 \bmod 6$$
$$X \equiv 2 \bmod 35$$

or is it merely a coincidence in this case that the reduced system is compatible? We will answer this question before too long.

To find the solutions of

$$X \equiv 7 \bmod 30$$
$$X \equiv 2 \bmod 35$$

we solve $30Y + 35Z = 7 - 2 = 5$. We find that $y = -1, z = 1$ is a solution to this equation, so we know from the proof of Lemma 6.7 that $35Z + 2 = 35(1) + 2 = 37$ is a simultaneous solution. Furthermore, the least common multiple of 30 and 35 is 210, so by Lemma 6.8 the set of all solutions of this last system is $\{x | x \equiv 37 \bmod 210\}$. But this is also the set of all solutions of the original system.

$$X \equiv -3 \bmod 10$$
$$X \equiv 1 \bmod 6$$
$$X \equiv 2 \bmod 35$$

Perhaps you now recognize that repetition of the foregoing procedure enough times can handle any compatible system, and what we then have is an algorithm for solving the system of congruences

$$\begin{aligned} X &\equiv c_1 \bmod m_1 \\ &\vdots \\ X &\equiv c_k \bmod m_k \end{aligned}$$

Algorithm.

(i) Look at the first two congruences. If they are not compatible, then there are no simultaneous solutions to the system. If they are compatible, use Lemmas 6.7 and 6.8 to replace the first two congruences by the single congruence

$$X \equiv x_0 \bmod m$$

where m is the least common multiple of m_1 and m_2, and x_0 is a simultaneous solution to the first two congruences.

(ii) Next look at the two congruences

$$X \equiv x_0 \bmod m$$
$$X \equiv c_3 \bmod m_3$$

and repeat (i) with these two congruences.

(iii) Continue.

Remember that if we combine Theorem 6.4 and the preceding algorithm, we can handle any system of congruences of the form

$$\begin{aligned} a_1 X &\equiv c_1 \bmod m_1 \\ &\vdots \\ a_k X &\equiv c_k \bmod m_k \end{aligned}$$

We are now ready to consider the converse of Theorem 6.6. Recall that in trying to solve the compatible system

$$X \equiv -3 \bmod 10$$
$$X \equiv 1 \bmod 6$$
$$X \equiv 2 \bmod 35$$

we replaced the system with the system of two congruences

$$X \equiv 7 \bmod 30$$
$$X \equiv 2 \bmod 35$$

This smaller system was again compatible and we asked: Is compatibility of the second system a lucky break in this example, or is it actually compatible because the original system was compatible? For this question, a general fact is useful.

Lemma 6.9. Let m_1, m_2, and n be positive integers and let m be the least common multiple of m_1 and m_2. Then (m, n) is a divisor of the least common multipe of (m_1, n) and (m_2, n).

Proof. Let us look at the prime decomposition of (n, m):

$$(n, m) = p_1^{a_1} p_2^{a_2} \cdots p_t^{a_t}$$

In order to show that (n, m) divides the least common multiple of (m_1, n) and (m_2, n), it is enough to show that this least common multiple is divisible by each of $p_i^{a_i}$ for $i = 1, \ldots, t$. To show this, however, we must only show that $p_i^{a_i} | (m_1, n)$ or $p_i^{a_i} | (m_2, n)$ and apply Exercise 3.B.14. Now we have $p_i^{a_i} | (m, n)$, so $p_i^{a_i} | m$ and $p_i^{a_i} | n$. Since m is the least common multiple of m_1 and m_2, either $p_i^{a_i} | m_1$ or $p_i^{a_i} | m_2$. In case $p_i^{a_i} | m_1$, then since $p_i^{a_i} | n$, we must have $p_i^{a_i} | (m_1, n)$. In case $p_i^{a_i} | m_2$, then we have $p_i^{a_i} | (m_2, n)$. In either case, we are done. ∎

Lemma 6.10. Suppose that the system of congruences

$$\begin{aligned} X &\equiv c_1 \bmod m_1 \\ X &\equiv c_2 \bmod m_2 \\ &\vdots \\ X &\equiv c_k \bmod m_k \end{aligned}$$

is compatible, and suppose that x_0 is a solution of the system

$$\begin{aligned} X &\equiv c_1 \bmod m_1 \\ X &\equiv c_2 \bmod m_2 \end{aligned}$$

If m is the least common multiple of m_1 and m_2, then the system

$$\begin{aligned} X &\equiv x_0 \bmod m \\ X &\equiv c_3 \bmod m_3 \\ &\vdots \\ X &\equiv c_k \bmod m_k \end{aligned}$$

is also compatible.

Proof. From the definition of compatibility we see that the system

$$X \equiv c_3 \bmod m_3$$
$$\vdots$$
$$X \equiv c_k \bmod m_k$$

is compatible because it is part of a larger compatible system, namely the one we started with. Thus we have only to check that for $i \geq 3$ $(m, m_i) | x_0 - c_i$. Now x_0 is a solution to the system

$$X \equiv c_1 \bmod m_1$$
$$X \equiv c_2 \bmod m_2$$

so that $m_1 | x_0 - c_1$ and $m_2 | x_0 - c_2$. However, $e_{1i} | m_1$, where $e_{1i} = (m_1, m_i)$ and $e_{2i} | m_2$. Thus $e_{1i} | x_0 - c_1$ and $e_{2i} | x_0 - c_2$. Since we also know that $e_{1i} | c_1 - c_i$ and $e_{2i} | c_2 - c_i$ we now deduce that e_{1i} divides $x_0 - c_i = (x_0 - c_1) + (c_1 - c_i)$ and e_{2i} divides $x_0 - c_i = (x_0 - c_2) + (c_2 - c_i)$. Thus we have $e_{1i} | (x_0 - c_i)$ and $e_{2i} | (x_0 - c_i)$. This of course means that the least common multiple of e_{1i} and e_{2i} divides $x_0 - c_i$. However, $e_{1i} = (m_1, m_i)$ and $e_{2i} = (m_2, m_i)$, so what we have is that the least common multiple of (m_1, m_i) and (m_2, m_i) divides $x_0 - c_i$. On the other hand, apply Lemma 6.9 with $n = m_i$ to conclude that (m, m_i) divides the least common multiple of (m_1, m_i) and (m_2, m_i). Thus $(m, m_i) | (x_0 - c_i)$, establishing compatibility. ∎

Finally, we arrive at our result.

Theorem 6.11. The system of congruences

$$X \equiv c_1 \bmod m_1$$
$$\vdots$$
$$X \equiv c_k \bmod m_k$$

has a simultaneous solution if and only if it is compatible.

Proof. We already know that if the system has a solution, then it is compatible. For the converse we use induction on the number k of congruences in the system. Clearly, a compatible system of only one equation has a solution. If $k = 2$, then again a compatible system has a solution, by Lemma 6.7. Thus we suppose that any compatible system of n congruences, $1 \leq n < k$, has a solution and we consider a compatible system with k congruences

$$X \equiv c_1 \bmod m_1$$
$$\vdots$$
$$X \equiv c_k \bmod m_k$$

Selecting the first two congruences from this system, we see that they form a compatible system and so a solution exists. Call it x_0. Now by Lemma 6.8 if m is the least common multiple of m_1 and m_2, then the solutions of the first pair coincide with the solutions of $X \equiv x_0 \bmod m$. Hence the solutions of the system of k congruences coincide with the solutions of the system of $k - 1$ congruences

$$\begin{aligned} X &\equiv x_0 \bmod m \\ X &\equiv c_3 \bmod m_3 \\ &\vdots \\ X &\equiv c_k \bmod m_k \end{aligned}$$

However by Lemma 6.10, this system is still compatible and hence by the induction hypothesis has a solution. Thus the original system of k congruences has a solution. ∎

6.3 The Congruence $X^2 \equiv a \bmod m$

In this section we will investigate the congruence

$$X^2 \equiv a \bmod m$$

Our attack will be to reduce the problem to the point where we only have to consider the case where m is a prime. As a first step, let us illustrate what happens when m is a product of two relatively prime integers.

EXAMPLE. Suppose that we wish to solve the congruence

$$X^2 \equiv 19 \bmod 425$$

We have $425 = 5^2 \cdot 17$ and $(5^2, 17) = 1$. We claim that it is enough to find solutions of the two congruences

$$X^2 \equiv 19 \bmod 25 \qquad \text{and} \qquad X^2 \equiv 19 \bmod 17$$

For instance, $(12)^2 \equiv 19 \bmod 25$ and $(6)^2 \equiv 19 \bmod 17$, as you can easily check. We will use these two numbers, 12 and 6, to construct a solution of

$$X^2 \equiv 19 \bmod 425$$

If $x \in \mathbb{Z}$ and $x \equiv 12 \bmod 25$, then $x^2 \equiv (12)^2 \equiv 19 \bmod 25$, and, similarly, any integer congruent to 6 modulo 17 is a solution of

$$X^2 \equiv 19 \bmod 17$$

Thus let us consider the system of congruences

$$X \equiv 12 \bmod 25$$
$$X \equiv 6 \bmod 17$$

Because $(25, 17) = 1$, the Chinese Remainder Theorem tells us that there is a simultaneous solution to this system. In fact, using the techniques of the previous section, you should have little difficulty in arriving at $x = 312$ as a simultaneous solution. However, any solution of the congruence

$$X \equiv 12 \bmod 25$$

is a solution of the congruence

$$X^2 \equiv 19 \bmod 25$$

At the same time any solution of

$$X \equiv 6 \bmod 17$$

is also a solution of

$$X^2 \equiv 19 \bmod 17$$

In particular, we have that $x = 312$ is the solution of both congruences

$$X^2 \equiv 19 \bmod 25 \qquad \text{and} \qquad X^2 \equiv 19 \bmod 17$$

Thus $25|(312)^2 - 19$ and $17|(312)^2 - 19$. However, $(25, 17) = 1$, so by Exercises 3.B.14 and 2.B.6, we must have $25 \cdot 17|(312)^2 - 19$. That is, 312 is a solution of the congruence

$$X^2 \equiv 19 \bmod 425$$

In what we have just done, we used the fact that $(12)^2 \equiv 19 \bmod 25$ and $6^2 \equiv 19 \bmod 17$. However, $X^2 \equiv 19 \bmod 25$, as well as $X^2 \equiv 19 \bmod 17$, have other solutions as well. By examining the numbers

$$1^2, 2^2, 3^2, \ldots, (24)^2 \text{ modulo } 25$$

we find that $X^2 \equiv 19 \bmod 25$ has two solutions, namely 12 and 13. Similarly, examining the squares from 1 to 16 modulo 17, we see that $X^2 \equiv 19 \bmod 17$ also has two solutions, 6 and 11. We have already used 12 and 6 to find a simultaneous solution, 312, to the system

$$X \equiv 12 \bmod 25$$
$$X \equiv 6 \bmod 17$$

which in turn was then a solution to $X^2 \equiv 19 \bmod 425$. We could also have solved the system

$$X \equiv 13 \bmod 25$$
$$X \equiv 6 \bmod 17$$

A simultaneous solution is $x = 363$ and, reasoning as before, 363 is also a solution of $X^2 \equiv 19 \bmod 425$. If we worked instead with

$$X \equiv 13 \bmod 25$$
$$X \equiv 11 \bmod 17$$

we would arrive at the solution $x = 113$. Finally, we could also take

$$X \equiv 12 \bmod 25$$
$$X \equiv 11 \bmod 17$$

and obtain $x = 62$ as a solution to $X^2 \equiv 19 \bmod 425$. Thus by looking at the two sets of solutions, $\{12, 13\}$ of $X^2 \equiv 19 \bmod 25$ and $\{6, 11\}$ of $X^2 \equiv 19 \bmod 17$ and taking all pairs of one from each, we obtained the four solutions 62, 113, 312, and 363 to the congruence

$$X^2 \equiv 19 \bmod 425$$

If you cared to check the numbers $1^2, 2^2, \ldots, (424)^2$, you would find that these are the solutions modulo 425.

It is now time to formalize this argument. The statement of the general result is involved, but remembering the example will help keep things in their proper place.

Theorem 6.12. Let $m > 0$ and $m = m_1 m_2$, where $(m_1, m_2) = 1$. Let $c_1, c_2, \ldots, c_r$ be all the distinct solutions modulo m_1 of the congruence

$$X^2 \equiv a \bmod m_1$$

and let $d_1, d_2, \ldots, d_s$ be all the distinct solutions modulo m_2 of the congruence

$$X^2 \equiv a \bmod m_2$$

For each pair of subscripts i, j where $1 \le i \le r$ and $1 \le j \le s$, let f_{ij} be a simultaneous solution of the system

$$X \equiv c_i \bmod m_1$$
$$X \equiv d_j \bmod m_2$$

Then

(i) f_{ij} is a solution of $X^2 \equiv a \bmod m$.
(ii) The f_{ij} are all noncongruent modulo m.
(iii) Any solution of $X^2 \equiv a \bmod m$ is congruent modulo m to some f_{ij}.

Proof. First note that the condition $(m_1, m_2) = 1$ enables us to invoke the Chinese Remainder Theorem and ensure ourselves that a solution f_{ij} exists to the system

$$X \equiv c_i \bmod m_1$$
$$X \equiv d_j \bmod m_2$$

(i) Now c_i is solution of the congruence $X^2 \equiv a \bmod m_1$, so that $c_i^2 \equiv a \bmod m_1$. Since $f_{ij} \equiv c_i \bmod m_1$ we have $(f_{ij})^2 \equiv a \bmod m_1$. Likewise, d_j satisfies $X^2 \equiv a \bmod m_2$, so $d_j^2 \equiv a \bmod m_2$. Hence $(f_{ij})^2 \equiv a \bmod m_2$, since $f_{ij} \equiv d_j \bmod m_2$. Therefore, we have $m_1 | (f_{ij})^2 - a$ and $m_2 | (f_{ij})^2 - a$. Since $(m_1, m_2) = 1$, this implies that $m_1 m_2 | (f_{ij})^2 - a$. Accordingly,

$$(f_{ij})^2 \equiv a \bmod m$$

so f_{ij} is a solution of the congruence $X^2 \equiv a \bmod m$.

(ii) We have to show that if $f_{ij} \equiv f_{kl} \bmod m$, then $i = k$ and $j = l$. Now $m_1 | m$ and $m | f_{ij} - f_{kl}$; hence

$$f_{ij} \equiv f_{kl} \bmod m_1$$

We know that $f_{ij} \equiv c_i \bmod m_1$ and $f_{kl} \equiv c_k \bmod m_1$ and thus $c_i \equiv c_k \bmod m_1$. Since we were assuming that $c_1, c_2, \ldots, c_r$ are all the distinct solutions modulo m_1 to $X^2 \equiv a \bmod m_1$, we must have $i = k$. Working with m_2, a parallel argument establishes that $j = l$.

(iii) Let f be a solution of $X^2 \equiv a \bmod m$. Then $f^2 \equiv a \bmod m$. As $m_1 | m$ and $m_2 | m$, we then have

$$f^2 \equiv a \bmod m_1 \qquad \text{and} \qquad f^2 \equiv a \bmod m_2$$

Thus for some i and j, where $1 \leq i \leq r$ and $1 \leq j \leq s$, we must have

$$f \equiv c_i \bmod m_1, \qquad f \equiv d_j \bmod m_2$$

Hence f is a simultaneous solution of the system

$$X \equiv c_i \bmod m_1$$
$$X \equiv d_j \bmod m_2$$

But $(m_1, m_2) = 1$, so their least common multiple is m. Consequently, by Lemma 6.8, all solutions of this system are described by the set $\{x \mid x \equiv f \bmod \mathrm{m}\}$. Thus $f_{ij} \equiv f \bmod m$, completing the proof. ∎

One obvious consequence of this result is that if $m = m_1 m_2$, where $(m_1, m_2) = 1$, then the number of distinct solutions modulo m of the congruence $X^2 \equiv a \bmod m$ is the product of the number of distinct solutions modulo m_1 and m_2, respectively, of $X^2 \equiv a \bmod m_1$ and $X^2 \equiv a \bmod m_2$. Induction easily gives the following.

Corollary 6.13. Let $m > 1$ with prime decomposition $m = p_1^{e_1} p_2^{e_2} \cdots p_k^{e_k}$. For each $i = 1, \ldots, k$, let r_i be the number of distinct solutions modulo $p_i^{e_i}$ of the congruence $X^2 \equiv a \bmod(p_i^{e_i})$. Then the number of distinct solutions modulo m of $X^2 \equiv a \bmod m$ is $r_1 r_2 \cdots r_k$.

You should keep in mind that it can very well happen that a congruence of the form $X^2 = a \bmod m$ may have no solutions. Such a congruence is

$$X^2 \equiv 2 \bmod 9$$

You may check directly that none of $1^2, 2^2, 3^2, \ldots, 8^2$ is congruent to 2 modulo 9.

What we have accomplished so far is to show that in order to solve $X^2 \equiv a \bmod m$, it is enough to solve the corresponding congruences

$$X^2 \equiv a \bmod m_1, \qquad X^2 \equiv a \bmod m_2$$

where $m = m_1 m_2$ and $(m_1, m_2) = 1$. An obvious choice for a splitting of m into relatively prime factors is the prime factorization of m. Therefore, if $m = p_1^{e_1} p_2^{e_2} \cdots p_k^{e_k}$, in order to solve $X^2 \equiv a \bmod m$ it is enough to solve each of the congruences

$$X^2 \equiv a \bmod p_i^{e_i}, \qquad 1 \le i \le k$$

Thus we are led to consider solutions of congruences of the form

$$X^2 \equiv a \bmod(p^e); \qquad p \text{ a prime}, \quad e > 0$$

We now wish to reduce the problem further and show that it is enough to be able to solve congruences of the form.

$$X^2 \equiv a \bmod p, \qquad p \text{ a prime}$$

The argument we are about to present works almost all the time. The exceptions occur if either $p = 2$ or if $p \mid a$. However, these two cases are essentially trivial and we will outline their treatment in the exercises at the end of the chapter. The main thrust of the ideas occurs in the case $p \nmid a$ and $p \neq 2$. In what follows we will have this as a blanket assumption.

You may have noticed that for $m > 0$ and $x \in \mathbb{Z}$, $x^2 \equiv (m - x)^2 \bmod m$. Of course, this is clear, since $m - x \equiv -x \bmod m$, so $(m - x)^2 \equiv (-x)^2 \equiv x^2 \bmod m$. If $m = p^e$, where p is an odd prime, we can say even more.

Lemma 6.14. Let $p > 2$ be a prime and $a \in \mathbb{Z}$ with $(a, p) = 1$. Then the congruence

$$X^2 \equiv a \bmod(p^e)$$

has either no solutions or exactly two solutions modulo p^e.

Proof. Suppose that there are solutions to the congruence and let x_0 be one of them. We will show that the only other solution modulo p^e is $p^e - x_0$. By the preceding remark, $p^e - x_0$ is a solution. Let x_1 be any solution. Then $x_1^2 \equiv a \bmod p^e$. As $x_0^2 \equiv a \bmod p^e$, we must have $x_0^2 \equiv x_1^2 \bmod(p^e)$. Thus $p^e | x_0^2 - x_1^2$. However, $x_0^2 - x_1^2 = (x_0 - x_1)(x_0 + x_1)$. We consider possible cases. If $p^e | x_0 + x_1$, then $x_1 \equiv -x_0 \equiv p^e - x_0 \bmod(p^e)$, while if $p^e | x_0 - x_1$, then $x_1 \equiv x_0 \bmod(p^e)$. In either of these two cases the conclusion is satisfied. What if $p^e \nmid (x_0 - x_1)$ and $p^e \nmid (x_0 + x_1)$? We claim that this case is impossible. Indeed, by examining the prime decomposition of $(x_0 - x_1)(x_0 + x_1)$, we see that since $p^e \nmid (x_0 - x_1)$, we must have $p | x_0 + x_1$; similarly, since $p^e \nmid (x_0 + x_1)$, $p | (x_0 - x_1)$. Therefore, p divides $(x_0 + x_1) - (x_0 - x_1) = 2x_1$. As $(p, 2) = 1$, we then have $p | x_1$. But x_1 is a solution of $X^2 \equiv a \bmod p^e$, so that $p^e | x_1^2 - a$. Since $p | x_1^2 - a$ and $p | x_1^2$, we arrive at the contradiction, $p | a$. ∎

A natural question to ask is whether the number of solutions of $X^2 \equiv a \bmod(p^e)$ is the same as the number of solutions of $X^2 \equiv a \bmod p$. For example, might it happen that $X^2 \equiv a \bmod 5$ has two solutions while $X^2 \equiv a \bmod 25$ has none? We have an answer.

Lemma 6.15. Suppose that $p > 2$ is a prime and $(a, p) = 1$. Then the number of solutions of the congruence $X^2 \equiv a \bmod(p^e)$ is the same as the number of solutions of the congruence $X^2 \equiv a \bmod(p^{e+1})$; that is, either both are 0 or both are 2.

Proof. Since we know by Lemma 6.14 that each has either no solutions or two solutions, it is enough to show that $X^2 \equiv a \bmod(p^e)$ is solvable if and only if $X^2 \equiv a \bmod(p^{e+1})$ is solvable. One direction is easy. If x_0 satisfies $X^2 \equiv a \bmod(p^{e+1})$, then $p^{e+1} | x_0^2 - a$, so $p^e | x_0^2 - a$; hence x_0 satisfies $X^2 \equiv a \bmod(p)$. The other direction is harder [observe: 3 is a solution of $X^2 \equiv 4 \bmod 5$ but not a solution of $X^2 \equiv 4 \bmod(5^2)$]. Let us suppose that

x_1 is a solution of $X^2 \equiv a \bmod(p^e)$. Then $p^e | x_1^2 - a$ and we then have $x_1^2 - a = k \cdot p^e$ for some $k \in \mathbb{Z}$. Note that $p \nmid x_1$, since $(p, \bar{a}) = 1$. Also, $p \neq 2$, so that $p \nmid 2x_1$. Hence $(2x_1, p) = 1$, so by Theorem 6.2 the congruence $(2x_1)Y \equiv -k \bmod p$ has a solution. Let y_0 be a solution. Then $p | 2x_1 y_0 + k$ and our claim is that $x_1 + y_0 p^e$ is a solution of $X^2 \equiv a \bmod(p^{e+1})$. We check this out:

$$(x_1 + y_0 p^e)^2 = x_1^2 + 2x_1 y_0 p^e + y_0^2 p^{2e} = (a + kp^e) + 2x_1 y_0 p^e + y_0^2 p^{2e}$$

Now $e > 0$, so $2e \geq e + 1$; hence $p^{e+1} | p^{2e}$. Thus $y_0^2 p^{2e} \equiv 0 \bmod(p^{e+1})$. Furthermore, $kp^e + 2x_1 y_0 p^e = (k + 2x_1 y_0)p^e$ and by our choice of y_0, $p | k + 2x_1 y_0$; thus $p^{e+1} | (k + 2x_1 y_0)p^e$, showing that $kp^e + 2x_1 y_0 p^e \equiv 0 \bmod(p^{e+1})$. Thus $(x_1 + y_0 p^e)^2 \equiv a \bmod(p^{e+1})$, as desired. ∎

Definition. Let p be a prime and $a \in \mathbb{Z}$. We shall say that a is a *quadratic residue mod p* if there is a solution to the congruence $X^2 \equiv a \bmod p$. If there are no solutions to the congruence $X^2 \equiv a \bmod p$, then we say that a is a *quadratic nonresidue mod p*.

Thus quadratic residues mod p are exactly those integers which have a square root modulo p. For example, 1 is a quadratic residue mod p for any prime p, since we can always solve $X^2 \equiv 1 \bmod p$. Also, -1 is a quadratic residue mod 5 since 2 is a solution of $X^2 \equiv -1 \bmod 5$; however, -1 is a quadratic nonresidue mod 7, as there are no solutions to the congruence $X^2 \equiv -1 \bmod 7$.

Corollary 6.16. If $p > 2$ is a prime and $(p, a) = 1$, then

$$X^2 \equiv a \bmod(p^e)$$

has exactly two solutions modulo p^e if a is a quadratic residue mod p, and no solutions if a is a quadratic nonresidue mod p.

Proof. By inductive use of Lemma 6.15, the number of solutions of $X^2 \equiv a \bmod(p^e)$ is the same as the number of solutions of $X^2 \equiv a \bmod p$. If a is a quadratic residue mod p, then by Lemma 6.14 this number is 2; otherwise, the number is 0. ∎

We now have reduced the problem of solving $X^2 \equiv a \bmod m$ to the problem of solving each of the congruences $X^2 \equiv a \bmod p_i$, $1 \leq i \leq k$, where $m = p_1^{e_1} p_2^{e_2} \cdots p_k^{e_k}$ (except in the special cases where either $p_i | a$ or $p_i = 2$ for some i).

As an example, we consider the congruence

$$X^2 \equiv 2 \bmod 106{,}558{,}057$$

Notice that 106,558,057 $= 17^3 \cdot 23^2 \cdot 41$. We first consider the congruences

$$X^2 \equiv 2 \bmod 17$$
$$X^2 \equiv 2 \bmod 23$$
$$X^2 \equiv 2 \bmod 41$$

By examining the numbers $1^2, 2^2, \ldots, (16)^2$ we see that $6^2 \equiv 2 \bmod 17$, so that 2 is a quadratic residue mod 17. Similar inspection shows that $5^2 \equiv 2 \bmod 23$ and that $17^2 \equiv 2 \bmod 41$. Thus by Corollary 6.16, each of the congruences

$$X^2 \equiv 2 \bmod(17^3)$$
$$X^2 \equiv 2 \bmod(23^2)$$
$$X^2 \equiv 2 \bmod(41)$$

has exactly two solutions modulo 17^3, 23^2, and 41, respectively. By Corollary 6.13, $X^2 \equiv 2 \bmod 106{,}558{,}057$ has $2 \cdot 2 \cdot 2 = 8$ distinct solutions modulo 106,558,057.

If we wish to find these eight solutions we must work harder. First, let us find a solution to $X^2 \equiv 2 \bmod 17^3$. We know that $6^2 \equiv 2 \bmod 17$ and in fact $6^2 - 2 = 2 \cdot 17$. Following the argument in the proof of Lemma 6.15, we find a solution of the congruence

$$2 \cdot 6Y \equiv -2 \bmod 17$$

Such a solution is $y = -3$. Then we know that $6 + (-3)17 = -45$ is a solution of $X^2 \equiv 2 \bmod(17^2)$. In fact, $(-45)^2 - 2 = 7 \cdot 17^2$. Thus we next solve

$$2 \cdot (-45)Y \equiv -7 \bmod 17$$

A solution is $y = 15$. Therefore, $-45 + 15 \cdot 17^2 = 4290$ is a solution of $X^2 \equiv 2 \bmod(17^3)$. Of course, there is another solution modulo 17^3 and for that one we can take -4290.

Next we consider $X^2 \equiv 2 \bmod(23)^2$. Since $5^2 \equiv 2 \bmod 23$ and $5^2 - 2 = 1 \cdot 23$, we solve the congruence

$$2 \cdot 5Y \equiv -1 \bmod 23$$

getting $y = -7$. Then $5 + (-7)23 = -156$ is a solution of $X^2 \equiv 2 \bmod(23)^2$, and for our other solution we take 156.

Finally, the solutions of $X^2 \equiv 2 \bmod 41$ are 17 and -17.

Now to find the eight solutions of the congruence

$$X^2 \equiv 2 \bmod 106{,}558{,}057$$

we must, in accordance with Theorem 6.12, find the solutions of the following eight systems of congruences:

1. $X \equiv 4290 \bmod(17^3)$
   $X \equiv 156 \bmod(23^2)$
   $X \equiv 17 \bmod(41)$
2. $X \equiv 4290 \bmod(17^3)$
   $X \equiv 156 \bmod(23^2)$
   $X \equiv -17 \bmod(41)$
3. $X \equiv 4290 \bmod(17^3)$
   $X \equiv -156 \bmod(23^2)$
   $X \equiv 17 \bmod(41)$
4. $X \equiv 4290 \bmod(17^3)$
   $X \equiv -156 \bmod(23^2)$
   $X \equiv -17 \bmod(41)$
5. $X \equiv -4290 \bmod(17^3)$
   $X \equiv 156 \bmod(23^2)$
   $X \equiv 17 \bmod(41)$
6. $X \equiv -4290 \bmod(17^3)$
   $X \equiv 156 \bmod(23^2)$
   $X \equiv -17 \bmod(41)$
7. $X \equiv -4290 \bmod(17^3)$
   $X \equiv -156 \bmod(23^2)$
   $X \equiv 17 \bmod(41)$
8. $X \equiv -4290 \bmod(17^3)$
   $X \equiv -156 \bmod(23^2)$
   $X \equiv -17 \bmod(41)$

This is a process we can handle. Admittedly, it is a lot of work, but it beats searching through the numbers $1^2, 2^2, 3^2, \ldots, (106{,}558{,}056)^2$ trying to find ones congruent to 2 modulo 106,558,056.

You will notice that the various steps in the example, while tedious, were entirely mechanical except for the first step. That is, we had to search for solutions of the congruences $X^2 \equiv 2 \bmod 17$, $X^2 \equiv 2 \bmod 23$, and $X^2 \equiv 2 \bmod 41$. Once we found these, a purely mechanical series of steps permitted us to find all solutions of $X^2 \equiv 2 \bmod(17^3)$, $X^2 \equiv 2 \bmod(23^2)$, and $X^2 \equiv 2 \bmod 41$. Finally, we have an algorithm for working out solutions to a system of congruences. Thus it is only in solving $X^2 \equiv a \bmod p$ that we have to stop and search; the remaining steps are routine. This raises the question of whether or not there is some algorithm for finding solutions to the congruence $X^2 \equiv a \bmod p$. At least, is there a process for determining whether or not a is a quadratic residue mod p? [Of course, we can always check out the squares $1^2, 2^2, \ldots, (p-1)^2$, but no one in their right mind would like to try this with the prime

$$p = 170{,}141{,}183{,}460{,}469{,}231{,}731{,}687{,}303{,}715{,}884{,}105{,}727$$

Would you care to try this method in order to determine whether or not 337,900,435,829,654 is a quadratic residue of this prime?]

In Chapter 9, we will develop a method for answering such questions. The heart of the method will be a theorem known as the *Quadratic Reciprocity*

Theorem, which has a rich history. Euler guessed that the theorem was true, but its proof was beyond even his genius. It was not until Carl Friedrich Gauss turned his astonishing powers to the problem that it was resolved. The theorem has many uses other than the one presented here. Gauss considered it such a lovely and intriguing result that he devoted his best efforts to it continually throughout his life. Eventually, he produced at least six different proofs of the theorem, one of which reduces the problem to that of considering what figures in Euclidean geometry can be constructed using a straightedge and compass. Many consider Gauss the greatest mathematician of all time; certainly the only other possible rivals for that title would be Archimedes and Newton. Gauss is not only responsible for the idea of congruences but he also formulated the concept of the telegraph!

We close this section by briefly mentioning the congruence

$$aX^2 + bX + c \equiv 0 \bmod m$$

We first point out that Theorem 6.12 and Corollary 6.13 can be applied to this situation without much change (Exercise 6.B.13). Thus it is enough to consider congruences of the form

$$aX^2 + bX + c \equiv 0 \bmod p^e; \qquad p \text{ a prime}, \quad e > 0$$

We will discuss how to proceed in the special case that $p \neq 2$ and $(a, p) = 1$.

Since $(a, p) = 1$, $(a, p^e) = 1$ and we know by Theorem 4.3 that there is a $d \in Z$ such that

$$da \equiv 1 \bmod p^e$$

Obviously, the solutions of

$$aX^2 + bX + c \equiv 0 \bmod p^e$$

are the same as for the congruence obtained by multiplying by d.

$$X^2 + (bd)X + (cd) \equiv 0 \bmod p^e$$

Therefore, we might as well assume from the beginning that $a \equiv 1 \bmod p$, so our congruence looks like

$$X^2 + (bd)X + (cd) \equiv 0 \bmod p^e$$

Furthermore, we may assume that b is even, because if b is odd, we can replace it with $b + p^e$ since $b + p^e \equiv b \bmod p^e$ and since $b + p^e$ is even (recall that we are assuming that $p \neq 2$).

By "completing the square," we have

$$\left(X + \frac{b}{2}\right)^2 + \left(c - \frac{b^2}{4}\right) = X^2 + bX + c$$

so our problem is to solve

$$\left(X + \frac{b}{2}\right)^2 + \left(c - \frac{b^2}{4}\right) \equiv 0 \bmod p^e$$

If we let $Y = X + \frac{b}{2}$ and $k = -\left(c - \frac{b^2}{4}\right)$, then we get

$$Y^2 \equiv k \bmod p^e$$

In this way, we see that the problem of solving

$$aX^2 + bX + c \equiv 0 \bmod p^e; \qquad (a, p) = 1, \quad p \neq 2$$

reduces to solving

$$Y^2 \equiv k \bmod p^e$$

and we have just discussed this type of congruence.

6.4 A General Theorem

We will conclude by providing a result on the number of solutions to a polynomial congruence modulo a prime p. We will need it when we return to quadratic congruences in Chapter 9, as well as in Chapter 8. This is a convenient place to include it. We first have two lemmas.

Lemma 6.17. Let $h(X) = a_n X^n + a_{n-1}X^{n-1} + \cdots + a_1 X + a_0$ be a polynomial with integral coefficients and let $c \in \mathbb{Z}$. Then there is a polynomial $g(X)$ with integral coefficients and an integer r such that

$$h(X) = (X - c)g(X) + r$$

Proof. We shall use induction on the highest power n of X occurring in $h(X)$, the degree of $h(X)$. If $n = 1$, then $h(X) = a_1 X + a_0$, so we can let $g(X) = a_1$ and $r = a_0 + a_1 c$. Then $(X - c)g(X) + r = (X - c)a_1 + (a_0 + a_1 c) = a_1 X + a_0 = h(X)$. Thus the result is true when $n = 1$. Hence assume that the conclusion holds for any polynomial having $n \leq k - 1$ and suppose that $h(X) = a_k X^k + a_{k-1}X^{k-1} + \cdots + a_1 X + a_0$. Consider $h(X) - a_k X^{k-1} \cdot (X - c)$. Since $h(X)$ and $a_k X^{k-1}(X - c)$ both have degree k, it is clear that the highest power of X occurring in $h(X) - a_k X^{k-1}(X - c)$ will be less than or equal to $k - 1$. By the induction assumption there is a polynomial $g_0(X)$ with integer coefficients and an integer r such that

$$h(X) - a_k X^{k-1}(X - c) = g_0(X)(X - c) + r$$

Thus

$$h(X) = [a_k X^{k-1} + g_0(X)](X - c) + r$$

and we take $g(X) = a_k X^{k-1} + g_0(X)$. ∎

Lemma 6.18. Let $m > 0$ and let $h(X) = a_n X^n + \cdots + a_1 X + a_0$, where $a_i \in \mathbb{Z}$, $0 \le i \le n$. If $c \in \mathbb{Z}$ is a solution of the congruence

$$h(X) \equiv 0 \bmod m$$

then there is a polynomial $g(X)$ with integer coefficients and an integer r such that $m \mid r$ and $h(X) = (X - c)g(X) + r$.

Proof. By the previous lemma we have $h(X) = (X - c)g(X) + r$, where $r \in \mathbb{Z}$ and $g(X)$ is a polynomial with integer coefficients. However, since c is a solution of $h(X) \equiv 0 \bmod m$, we must have

$$0 \equiv h(c) \equiv (c - c)g(c) + r \equiv r \bmod m$$

and hence $m \mid r$. ∎

Theorem 6.19. Let p be a prime and let $h(X) = a_n X^n + \cdots + a_1 X + a_0$ be a polynomial with integer coefficients where $n \ge 1$. If $(a_n, p) = 1$, then there are at most n solutions modulo p to the congruence

$$h(X) = a_n X^n + \cdots + a_1 X + a_0 \equiv 0 \bmod p$$

Proof. We again resort to inducting on n. If $n = 1$, we have the congruence

$$a_1 X + a_0 \equiv 0 \bmod p, \qquad \text{where } (a_1, p) = 1$$

That is,

$$a_1 X \equiv -a_0 \bmod p$$

As we know well by now, there is a unique solution modulo p. Thus we suppose that the result is true whenever $1 \le n \le k - 1$ and suppose that $h(X) = a_k X^k + \cdots + a_1 X + a_0$ with $(a_k, p) = 1$. Let us assume that the result is false and that there are at least $k + 1$ solutions modulo p to the congruence $h(X) \equiv 0 \bmod p$. Thus there are integers $c_1, c_2, \ldots, c_{k+1}$ such that $h(c_i) \equiv 0 \bmod p$, $1 \le i \le k + 1$, and $c_i \not\equiv c_j \bmod p$ whenever $i \ne j$. We apply Lemma 6.18 to c_{k+1} to obtain a polynomial $g(X)$ with integer coefficients and an integer r such that $h(X) = (X - c_{k+1})g(X) + r$ and $p \mid r$. Now for any $1 \le i \le k$, we have

$$0 \equiv h(c_i) \equiv (c_i - c_{k+1})g(c_i) + r \equiv (c_i - c_{k+1})g(c_i) \bmod p$$

because $r \equiv 0 \bmod p$. Thus

$$(c_i - c_{k+1})g(c_i) \equiv 0 \bmod p$$

which means that either $c_i - c_{k+1} \equiv 0 \bmod p$ or $g(c_i) \equiv 0 \bmod p$. However, $i \neq k + 1$, so that $c_i - c_{k+1} \not\equiv 0 \bmod p$. It follows then that for each $i = 1, 2, \ldots, k$, we have $g(c_i) \equiv 0 \bmod p$, so the congruence

$$g(X) \equiv 0 \bmod p$$

has at least k distinct solutions mod p. But $h(X) = (X - c_{k+1})g(X) + r$, so the highest power of X occurring in $g(X)$ is $k - 1$, and the coefficient of X^{k+1} is a_k. Since $(a_k, p) = 1$, the induction hypothesis tells us that there are at most $k - 1$ distinct solutions modulo p to the congruence

$$g(X) \equiv 0 \bmod p$$

This is the desired contradiction. Thus the result is also true for k, completing the proof. ∎

EXERCISES 6.A

1. Determine how many solutions exist modulo m for each of the following congruences.
   (a) $3X \equiv 11 \bmod 17$ (b) $3X \equiv 11 \bmod 15$
   (c) $3X \equiv 12 \bmod 15$ (d) $-3X \equiv 12 \bmod 15$
   (e) $8X \equiv -24 \bmod 12$ (f) $12X \equiv 68 \bmod 40$
   (g) $49X \equiv 21 \bmod 35$ (h) $121X \equiv 185 \bmod 77$
2. Find a solution to each of the following congruences.
   (a) $3X \equiv 2 \bmod 5$ (b) $7X \equiv 5 \bmod 12$
   (c) $14X \equiv 72 \bmod 11$ (d) $25X \equiv 10 \bmod 30$
   (e) $42X = 21 \bmod 35$ (f) $18X \equiv 24 \bmod 30$
   (g) $105X \equiv 35 \bmod 70$ (h) $71X \equiv 1543 \bmod 73$
3. Find all solutions modulo m for each of the congruences in Exercise 6.A.2.
4. Find all solutions modulo m for each of the following congruences.
   (a) $5X \equiv 5 \bmod 5$ (b) $5X \equiv 5 \bmod 10$
   (c) $5X \equiv 5 \bmod 15$
5. Let $d = (a, m)$ and assume that $d \mid c$. Show that the congruence $aX \equiv c \bmod m$ and the congruence $(a/d)X \equiv (c/d) \bmod (m/d)$ have exactly the same solutions.

6. Reduce each of the following systems of equations to a system of the type

$$x \equiv c_1 \bmod m_1$$
$$\vdots$$
$$x \equiv c_n \bmod m_n$$

having the same solution set, if it is possible to do so (i.e., if Proposition 6.3 is happy).

(a) $3X \equiv 3 \bmod 30$
$4X \equiv 2 \bmod 42$

(b) $15X \equiv 35 \bmod 10$
$14X \equiv 30 \bmod 31$

(c) $18X \equiv 6 \bmod 24$
$45X \equiv 36 \bmod 54$

(d) $10X \equiv 6 \bmod 12$
$6X \equiv 3 \bmod 27$

(e) $12X \equiv 24 \bmod 45$
$6X \equiv 2 \bmod 20$

(f) $9X \equiv 6 \bmod 12$
$5X \equiv 10 \bmod 15$
$2X \equiv 4 \bmod 120$

(g) $25X \equiv 10 \bmod 30$
$9X \equiv 18 \bmod 30$
$14X \equiv 8 \bmod 30$

7. In Exercise 6.A.6(g) discuss why the answer is *not*

$$X \equiv 4 \bmod 30$$
$$X \equiv 2 \bmod 30$$
$$X \equiv 7 \bmod 30$$

even though 4, 2, and 7 are, respectively, solutions to the separate congruences.

8. Determine which of the following are compatible systems.

(a) $X \equiv 2 \bmod 12$
$X \equiv 8 \bmod 16$

(b) $X \equiv 4 \bmod 21$
$X \equiv 18 \bmod 35$
$X \equiv 8 \bmod 15$

(c) $X \equiv 4 \bmod 21$
$X \equiv 18 \bmod 35$
$X \equiv 13 \bmod 15$

9. Which of the answers obtained in Exercise 6.A.6 are compatible systems?

10. Consider the system

$$X \equiv 2 \bmod 12$$
$$X \equiv 8 \bmod 30$$
$$X \equiv c \bmod 20$$

Notice that the first pair of congruences is compatible. Find a value of c which makes all three congruences compatible.

11. Use the process given in the proof of Lemma 6.7 to find a solution for each of the following systems (if possible).

(a) $X \equiv 13 \bmod 33$
$X \equiv 35 \bmod 35$

(b) $X \equiv 6 \bmod 10$
$X \equiv 11 \bmod 25$

(c) $X \equiv -1 \bmod 4$
$X \equiv 2 \bmod 6$

(d) $X \equiv 3 \bmod 15$
$X \equiv 9 \bmod 18$

12. Find all solutions to each of systems (a) through (d) of Exercise 6.A.11. Your answers should be expressed in the form $X \equiv a \bmod m$.

13. Solve each of the following systems of congruences.

(a) $X \equiv 6 \bmod 10$
$X \equiv 11 \bmod 25$
$X \equiv 16 \bmod 15$

(b) $X \equiv 13 \bmod 33$
$X \equiv 35 \bmod 35$
$X \equiv 24 \bmod 77$

(c) $X \equiv 3 \bmod 15$
$X \equiv 9 \bmod 18$
$X \equiv 13 \bmod 10$

(d) $X \equiv 12 \bmod 14$
$X \equiv 8 \bmod 10$
$X \equiv 33 \bmod 35$

(e) $X \equiv 11 \bmod 30$
$X \equiv 23 \bmod 42$
$X \equiv 51 \bmod 70$
$X \equiv 86 \bmod 105$

14. Verify Lemma 6.9 for $m_1 = 30$, $m_2 = 70$, and $n = 231$.

15. Solve:

(a) $X \equiv 18 \bmod 21$
$X \equiv 12 \bmod 15$
$X \equiv 25 \bmod 35$

(b) $2X \equiv 2 \bmod 8$
$3X \equiv 9 \bmod 10$
$5X \equiv 2 \bmod 7$

(c) $2X \equiv 4 \bmod 14$
$7X \equiv 3 \bmod 10$
$6X \equiv 5 \bmod 9$

(d) $10X \equiv 22 \bmod 24$
$9X \equiv 15 \bmod 84$
$15X \equiv 123 \bmod 126$

(e) $3X \equiv 12 \bmod 18$
$2X \equiv -8 \bmod 20$
$3X \equiv 42 \bmod 45$

(f) $8X \equiv 28 \bmod 36$
$5X \equiv 20 \bmod 50$
$7X \equiv 5 \bmod 15$

(g) $3X \equiv -1 \bmod 8$
$5X \equiv 5 \bmod 10$
$6X \equiv 6 \bmod 15$
$3X \equiv 9 \bmod 18$

(h) $5X \equiv -1 \bmod 12$
$3X \equiv -15 \bmod 54$
$3X \equiv 11 \bmod 27$

16. By inspection, find all solutions modulo 7 of $X^2 \equiv 4 \bmod 7$ and all solutions modulo 9 of $X^2 \equiv 4 \bmod 9$. Then use Theorem 6.12 to find all solutions modulo 63 of $X^2 \equiv 4 \bmod 63$.

17. By inspection, find all solutions modulo 8 of $X^2 \equiv 1 \bmod 8$ and all solutions modulo 5 of $X^2 \equiv 1 \bmod 5$. Then use Theorem 6.12 to find all solutions modulo 40 of $X^2 \equiv 1 \bmod 40$.

18. Proceeding as in Exercises 6.A.16 and 17, find all solutions for each of the following congruences.
(a) $X^2 \equiv 1 \bmod 45$ (b) $X^2 \equiv 3 \bmod 40$
(c) $X^2 \equiv 3 \bmod 143$ (d) $X^2 \equiv 4 \bmod 385$

19. Let $m = p_1^{e_1} p_2^{e_2} \cdots p_t^{e_t}$ be the prime decomposition of the *odd* integer m. Use Corollary 6.13 and Lemma 6.14 to show that if $(a, m) = 1$, then the congruence $X^2 \equiv a \bmod m$ either has no solutions or else has 2^t solutions modulo m.

20. Find all quadratic residues mod p for the following values of the prime p: 5, 7, 11, 13, 17.

21. Use the process given in the proof of Lemma 6.15 to find a solution of $X^2 \equiv 2 \bmod 49$. Thus, first find a solution of $X^2 \equiv 2 \bmod 7$, say u. Let $k = (u^2 - 2)/7$. Letting v be a solution of $(2u)Y \equiv -k \bmod 7$, verify that $u + 7v$ is a solution of $X^2 \equiv 2 \bmod 49$.

22. Proceeding as in Exercise 6.A.21, find a solution of the congruence $X^2 \equiv 3 \bmod 121$.

23. Using your solution of $X^2 \equiv 2 \bmod 49$ found in Exercise 6.A.21, find a solution of the congruence $X^2 \equiv 3 \bmod 343$ using the process in Lemma 6.15. (*Note:* $343 = 7^3$.)

24. Proceed as in Exercise 6.A.21 to find a solution of $X^2 \equiv 3 \bmod 11^3$.

25. How many distinct solutions exist modulo m for each of the following congruences?
(a) $X^2 \equiv 1 \bmod 63$ (b) $X^2 \equiv 1 \bmod 385$
(c) $X^2 \equiv 3 \bmod 1575$

26. Solve each of the congruences in Exercise 6.A.25.

27. Rewrite the following congruences in the form $Y^2 \equiv k \bmod p$.
(a) $4X^2 + 3X + 5 \equiv 0 \bmod 7$ (b) $3X^2 + 3X + 5 \equiv 0 \bmod 7$
(c) $17X^2 - 22X + 9 \equiv 0 \bmod 11$
(d) $-13X^2 + 73X + 19 \equiv 0 \bmod 11$

28. Solve the following congruences.
(a) $X^2 + 2X + 6 \equiv 0 \bmod 49$ (b) $X^2 + X + 9 \equiv 0 \bmod 49$
(c) $X^2 - X + 1 \equiv 0 \bmod 49$

29. For $h(X) = 7X^4 - 2X^3 - 5X^2 + 4X + 4$ and $c = 2$, find the $g(X)$ and r that satisfy Lemma 6.17.

30. With $h(X)$ and c as in Exercise 6.A.29, show that c is a solution of $h(X) \equiv 0 \bmod 11$. Then show that for $m = 11$, Lemma 6.18 is true.

31. Find all solutions of $X^3 - 3X^2 + 3X - 23 \equiv 0 \bmod 7$ and verify that Theorem 6.19 is satisfied.

32. In discussing $X^2 \equiv 2 \bmod 106{,}558{,}057$, we reduced the problem to solving eight systems of simultaneous congruences. Argue that we actually only must solve four of them. [*Hint:* Compare system (1) and system (8), for instance.]

EXERCISES 6.B

1. Describe all solutions of the congruence

$$pX \equiv p \bmod kp$$

where p is a prime and $k = 1, 2, \ldots$. (*Hint:* Look at Exercise 6.A.4.)

2. In exercise 6.A.10 show that the c's that make the system compatible are precisely the solutions to the system

$$c \equiv 2 \bmod 4$$
$$c \equiv 8 \bmod 10$$

3. Consider the system of congruences

$$X \equiv c_1 \bmod m_1$$
$$X \equiv c_2 \bmod m_2$$
$$X \equiv c_3 \bmod m_3$$

Assume that c_1 is given. Describe how c_2 can be chosen so that the first two equations are compatible. Now describe how c_3 can be chosen so as to make the system compatible.

4. Generalize Exercise 6.B.3 to a system of $n \geq 3$ congruences.

5. Consider the system of congruences

$$X \equiv c_1 \bmod m_1$$
$$X \equiv c_2 \bmod m_2$$
$$\vdots$$
$$X \equiv c_k \bmod m_k$$

where $(m_i, m_j) = 1$ for $i \neq j$. Let $M_i = \prod_{i \neq j} m_j$ and let u_i be a solution to the congruence $M_i Y \equiv 1 \bmod m_i$ for $i = 1, \ldots, k$. Show $c = c_1 M_1 u_1 + c_2 M_2 u_2 + \cdots + c_k M_k u_k$ is a solution of the system.

The next sequence of Exercises (6 through 9) involves the solutions of the congruence $X^2 \equiv a \bmod p^e$, where $p \mid a$. We let d be the largest integer such that $p^d \mid a$. Thus $a = p^d r$ where $(p, r) = 1$; e, d, and r will be referred to in the exercises.

6. Let f be the smallest integer such that $e/2 \leq f$. Show that x_0 is a solution of $X^2 \equiv 0 \bmod p^e$ if and only if $p^f \mid x_0$.

7. If $d \geq e$, then $a \equiv 0 \bmod p^e$, so we may use Exercise 6.B.6 to solve the congruence $X^2 \equiv a \bmod p^e$. Thus assume that $1 \leq d < e$. Suppose that

x_0 is a solution of the congruence and that $x_0 = p^h s$, where $(s, p) = 1$. Show that $2h \geq d$. (*Hint*: From the fact that $x_0^2 \equiv a \bmod p^e$, deduce that $p^d | x_0^2$.)

8. In Exercise 6.B.7 show that in fact $d = 2h$. [*Hint*: If $d < 2h$, deduce that $p^e \nmid (x_0^2 - a)$, using that $d < e$.] Conclude that if d is odd, then the congruence has no solutions.

9. Knowing that $d = 2h$, show that x_0 is a solution of the congruence if and only if x_0 is of the form $x_0 = p^h s$, where $(s, p) = 1$ and s is a solution of the congruence $X^2 \equiv r \bmod p^{e-2h}$. (*Note*: Since $p \nmid r$, this reduces the original problem with $p | a$ to the case treated in the text if $p \neq 2$.)

The next sequence of Exercises (10 through 12) involves the solutions of the congruence $X^2 \equiv a \bmod 2^e$. If $2 | a$, then Exercises 6.B.6 through 9 apply, so we will assume that a is odd. Further the cases $e = 1$ or $e = 2$ are easily solved by inspection; hence we will assume that $e \geq 3$. Let $S = \{1, 3, 5, \ldots, 2^{e-2} - 1\}$.

10. Show that if x and y are distinct members of the set S, then $x^2 \not\equiv y^2 \bmod 2^e$. [*Hint*: One of $x + y$ and $x - y$ is not divisible by 4. Thus if $2^e | x^2 - y^2$, deduce that $2^{e-1} | (x + y)$ or $2^{e-1} | (x - y)$.]

11. Given $c \in S$, show that the following hold:

(a) c, $2^{e-1} - c$, $2^{e-1} + c$, and $2^e - c$ are all solutions of the congruence $X^2 \equiv c^2 \bmod 2^e$;

(b) The four integers in (a) are distinct and lie between 1 and 2^e.

(c) Any solution between 1 and 2^e of $X^2 \equiv c^2 \bmod 2^e$ is one of the four integers in (a). (*Hint*: If y is a solution, deduce that $2^{e-1} | y + c$ or $2^{e-1} | y - c$.)

12. For a odd show that $X^2 \equiv a \bmod 2^e$ has no solutions unless $a \equiv c^2 \bmod 2^e$ for some $c \in S$, in which case it has four solutions modulo 2^e.

13. State and prove analogues of Theorem 6.12 and Corollary 6.13 for the congruence $aX^2 + bX + c \equiv 0 \bmod m$.

14. Suppose that the congruence $X^2 + bX + c \equiv 0 \bmod p$, where p is a prime, has two incongruent solutions u and v modulo p. Let $u^2 + bu + c = kp$ and $v^2 + bv + c = lp$. Show that $(2u + b)Y \equiv -k \bmod p$ and $(2v + b)Y \equiv -l \bmod p$ each has one solution modulo p, say y_1 and y_2, respectively. Show that $u + y_1 p$ and $u + y_2 p$ are distinct solutions modulo p^2 of $X^2 + bX + c \equiv 0 \bmod p^2$. [Hint: In the next exercise you will show that $b \equiv -(u + v) \bmod p$ and $c \equiv uv \bmod p$].

15. Prove the fact quoted in the hint for Exercise 6.B.14. [*Hint*: Observe that $(X^2 + bX + c) - (X - u)(X - v) = [(b + u + v)X + (c - uv)]$. Show that modulo p, this linear expression has both u and v as solutions.]

7

Sums of Squares

WE will be studying some classical problems in number theory in this chapter. It seems fitting, therefore, to begin with a brief outline of the development and origins of number theory, particularly as it relates to these problems.

The beginnings can be traced back to the Babylonians, who were good mathematicians. Archaeological work done in the 1940s on Babylonian tablets dating between 1900 B.C. and 1600 B.C. substantiate this statement and in fact support the claim that they were much better than was previously believed. An example of their mathematics is that they were aware of and made use of the relationship existing among sides of a right triangle (the sum of the squares of the sides is the square of the hypotenuse). The relationship is universally recognized as the Pythagorean theorem; but Pythagoras was not born until 570 B.C.

A characteristic of Babylonian mathematics was that it was very much geared to applications; it dealt with problems of surveying, architecture, and finance. Nevertheless, some theoretical work with a strong algebraic (i.e., numerical) flavor existed. It appears that some of this flavor influenced Greek mathematics, which no doubt inherited a good bit from Babylon. The Greeks made two significant adjustments. First, they concentrated on geometry rather than algebra. Second, and perhaps more important from the mathematical perspective, they conceived of mathematics as a logical deduction system with certain ground rules, or axioms, from which other results could be obtained. In contrast, mathematics to the Babylonians was a collection of facts, sometimes interrelated; however, little perception existed as to which were fundamental facts and which were consequences of fundamental facts. This should not be interpreted as a negative appraisal of Babylonian mathematics: their mathematics sufficed for the everyday needs of their civilization. A modern-day parallel has been cited by Richard Feynman, the distinguished Nobel laureate, physicist, and lecturer, who has likened modern physics to ancient Babylonian mathematics.* Although some physical theories are

* See R. Feynman, *The Character of Physical Law* (Cambridge, Mass.: The MIT Press, 1967).

obviously more basic than others, physicists do not have a given set of fundamental rules from which all else follows; still physics is a thriving, lively area of study with much to offer our civilization.

One of the earliest and most influential of the Greek mathematicians was Pythagoras. He traveled in Babylonia and Egypt and these visits probably served as the foundation for his mathematical knowledge. Pythagoras established his school in southern Italy, a school that survived for a century. It is difficult now to distinguish the work of Pythagoras from that of his students, since the students attributed much of what they did to Pythagoras. The Pythagoreans did show that $\sqrt{2}$ is not a rational number; that is, it is impossible to obtain the $\sqrt{2}$ by dividing one integer by another. This fact apparently disturbed them profoundly, since they had believed that all numbers evolved from the integers. As it turns out, they were correct in their belief to some extent. However, the methods for obtaining all numbers from the integers were very subtle, involving the concept of limit as developed by Newton, Dedekind, and others. In another vein, the Pythagoreans argued that the Earth was a sphere. Proof: During an eclipse of the moon, the Earth could be seen to cast a circular shadow. Note that there is some deep thinking involved in this simple argument—planetary orbits, for example.

Much later in Greek history, when their science was generally in decline, a return to algebra took place. The leader of this movement was Diophantos, who lived in Alexandria sometime during the second or third century A.D. Not much more is known about Diophantos' personal life, except for some information contained in a book entitled *Palatine Anthology*, a collection of short poems ranging over a variety of subjects. Among these are some mathematical problems in epigrammatic form, one of which reads:

> Here you see the tomb containing the remains of Diophantos, it is remarkable: Artfully it tells the measure of his life. The sixth part of his life God granted him for his youth. After a twelfth more his cheeks were bearded. After an additional seventh he kindled the light of marriage, and in the fifth year he accepted a son. Elas, a dear but unfortunate child, half of his father he was and this was also the span a cruel fate granted it, and he consoled his grief in the remaining four years of his life. By this devise of numbers, tell us the extent of his life.*

Thus Diophantos' entire biography constitutes a homework problem!

* From *Number Theory and Its History* by Oystein Ore. Copyright 1948, McGraw-Hill Book Company, Inc. Used by permission of McGraw-Hill Book Company.

In sharp contrast, Diophantos' professional work is well known. He studied equations that had integral solutions, for instance, the equation $aX + bY = c$ studied in Chapter 2 and the equations $X^2 + Y^2 = Z^2$, $X^2 + Y^2 = n$, and $X^2 + Y^2 + Z^2 + W^2 = n$, which we will study in this chapter. In his honor, equations of this type are called Diophantine equations. Diophantos was also a pioneer in the use of mathematical symbols rather than words in doing mathematics. What was perhaps most important for the future of number theory, as well as his fame, was that Diophantos wrote books. Unfortunately, some were lost, but portions of one, *Arithmetics*, survived. In the exercises we will reproduce some of the problems contained in this book.

The Middle Ages was a bleak period for intellectual pursuit in general, and mathematics was a victim of this time period. Still *Arithmetics* survived. In the early eleventh century, the Arab mathematician Al-Karkhi of Baghdad wrote a text, named after the Grand Vizier of Baghdad, which was entitled *Al-Fakhri*. The book not only contained sections that were copied verbatim from *Arithmetics* but also further developed the techniques of Diophantos. In the early thirteenth century, Leonardo Pisano wrote *Liber Abaci* and copied portions of *Al-Fakhri*. The Hindus also made use of *Arithmetics*, further attesting to the widespread use of the work.

With the dawning of the Renaissance, Greek mathematical manuscripts made their way into the hands of European scholars, who were amazed to see how far ahead of them the Greeks had been. In 1575, a German mathematician, Holtzman (who assumed the Greek name Xylander), published the first printed edition of *Arithmetics*. In the introduction he relates the profound effect the manuscript had on him:

> As for . . . algebra . . . I had not only acquired command of the subject, but also had advanced to the point of adding, giving variety, and in places even of making corrections to what such great and devoted teachers . . . had written about it. In short I came to believe that in Arithmetic and Logistic "I was somebody." . . . But when I first came upon the work of Diophantos, his method and reasoning so overwhelmed me that I scarcely knew whether to think of my former self with pity or with laughter . . . .*

In 1621, a better edition of *Arithmetics* was published by Bachet de Meziriac, and a copy of this edition came into the hands of Pierre Fermat. It is almost a certainty that this text turned Fermat's attention to the theory of numbers. He read it carefully, using the margins of his copy to make notes. One such

* From *Number Theory and Its History* by Oystein Ore. Copyright 1948, McGraw-Hill Book Company, Inc. Used by permission of McGraw-Hill Book Company.

note concerned the equation $a^n + b^n = c^n$, $n > 2$, which we mentioned in Chapter 4. Two other notes dealt with an important special case of the equation $X^2 + Y^2 = n$ and the equation $X^2 + Y^2 + Z^2 + W^2 = n$; both of these equations will be treated in this chapter. Fermat usually stated only the result, without indicating the proof. This might be considered unfortunate, except that it did produce a stimulation of interest in number theory. Later mathematicians set about to supply missing proofs. Among these were Euler (Swiss), Lagrange and Dirichlet (French), and Kummer and Gauss (German). As we will see, Euler and Lagrange did succeed in supplying proofs to two of Fermat's statements. Kummer attempted to prove " Fermat's Last Theorem "; in 1843 he thought he had such a proof· However, Dirichlet, who had put some effort into the problem himself, quickly found an error.* Kummer's attempts at patching the flaw failed. But in his attempts, he gave birth to the theory of ideals and the modern subject now known as algebraic number theory (a big brother of elementary number theory). Meanwhile, the equation $X^2 + Y^2 + Z^2 + W^2 = n$ gave rise to Waring's Problem, which we will discuss at the end of this chapter. The investigation of this problem has led deeply into the modern subject known as analytic number theory (a big sister to elementary number theory), which uses techniques similar to calculus.

No history of mathematics, regardless of how brief, is complete without mentioning Gauss. This giant profoundly affected most of mathematics (and some physics and astronomy). Professor Irving Kaplansky once said of Gauss: " When I read the works of Mathematicians like Euler or Lagrange, I think ' How clever they were! If only they had had a chance to go to a good graduate school.' But when I read Gauss there is no such feeling. He was a good graduate school all by himself."† Clearly, number theory was dear to Gauss' heart. He called it " the queen of mathematics." In 1801, Gauss published *Disquisitiones Arithmeticae*, quite possibly the most important mathematics text ever produced. In it he mentioned the number theoretic work of Euclid, Diophantos, Fermat, Euler, Lagrange, and Legendre. He then introduced congruences and used them to cast elementary number theory in its present form.

* Recently, some doubt has been cast on whether Kummer's error concerned Fermat's Last Theorem, and if it was Dedekind who found it. H. M. Edwards, "The Background of Kummer's Proof of Fermat's Last Theorem For Regular Primes," *Archive for History of Exact Sciences*, Vol. 14 (1975), pp. 219–236.

† From a lecture entitled "Disquisitiones Arithmeticae: How to Pronounce It and What It Says," circa 1970 at the University of Chicago.

We hope that these brief comments provide some small idea of the historic significance of what we are about to see. The principal source for these thoughts is the delightful text *Number Theory and Its History* by Oystein Ore (New York: McGraw-Hill Book Company, 1948). We encourage you to read it.

7.1 Pythagorean Triples

Triples of integers (a, b, c) such that $a^2 + b^2 = c^2$ are the object of study in this section. Of course, we always have $0^2 + c^2 = c^2$ and $c^2 + 0^2 = c^2$, but this situation is too dull to merit interest. Thus we insist that both a and b be nonzero. Also, if $a^2 + b^2 = c^2$, then $(-a)^2 + (-b)^2 = c^2$, and so forth. We have no interest in such variations, so we further insist that a, b, and c are positive integers. Recalling the Pythagorean Theorem, we see that from a geometrical viewpoint we are considering right triangles whose sides are of integral length.

Definition. A *Pythagorean triple* is a triple of positive integers (a, b, c) such that $a^2 + b^2 = c^2$. If (a, b, c) is a Pythagorean triple, then each of a and b is said to be a *leg* of the triple.

Since $3^2 + 4^2 = 5^2$, we see that $(3, 4, 5)$ is a Pythagorean triple whose legs are 3 and 4. Other triples are $(6, 8, 10)$ and $(5, 12, 13)$, which you can easily check. Another Pythagorean triple is $(4961, 6480, 8161)$. This one, along with many other triples involving large numbers, was known to the Babylonians. Pythagoras is credited with noting that if n is a positive integer, then $(2n + 1, 2n^2 + 2n, 2n^2 + 2n + 1)$ is a Pythagorean triple. Simple algebraic calculations bear this out. Lest you think that Pythagorean triples are rare, we start with an easy result which should dispel that belief. (Also see Exercise 7.B.8.)

Theorem 7.1. For each integer $a > 2$ there exist integers b and c such that (a, b, c) is a Pythagorean triple.

Proof. Suppose first that a is even. Then $2 \mid a$, so $4 \mid a^2$; hence $(a^2 - 4)/4$ and $(a^2 + 4)/4$ are both integers. Easy calculations show that if $b = (a^2 - 4)/4$ and $c = (a^2 + 4)/4$, then (a, b, c) is a Pythagorean triple. If a is odd, then $a = 2n + 1$ and by Pythagoras' observation, $(2n + 1, 2n^2 + 2n, 2n^2 + 2n + 1)$ is a Pythagorean triple. ∎

The constructions given in this theorem do not provide all Pythagorean triples. A complete solution to the problem of describing all Pythagorean triples has been known for some time. For instance, Fermat knew a solution.

The solution that we will present is not the one that Fermat was familiar with, however. It is a very simple argument that appears to be almost brand new. We take it from the article "Pythagorean Triples: A New, Easy-to-Derive Formula with Some Geometric Applications," by Andrea Rothbart and Bruce Paulsell, which appeared in *Mathematics Teacher*, Vol. 67 (March, 1974), pp. 215–218. The next theorem therefore seems to be a new result about an ancient subject.

Theorem 7.2. A triple (a, b, c) is a Pythagorean triple if and only if there exist integers u and v which satisfy the following conditions:

(i) $u > v > 0$.
(ii) $u \equiv v \bmod 2$.
(iii) uv is a perfect square.
(iv) $a = \sqrt{uv}$, $b = (u - v)/2$, $c = (u + v)/2$.

Proof. First suppose that u and v satisfy conditions (i) through (iv). Then (i) and (iv) imply that a, b, and c are positive. Also, (iii) and (iv) yield that a is an integer, while (ii) and (iv) yield that b and c are integers. (Note that $u \equiv v \bmod 2$ is simply a short way of saying that u and v are both even or both odd.) Finally, from (iv),

$$a^2 + b^2 = (\sqrt{uv})^2 + [(u - v)/2]^2 = \tfrac{1}{4}(u^2 + 2uv + v^2) = [(u + v)/2]^2 = c^2$$

Thus (a, b, c) is a Pythagorean triple.

Conversely, assume that (a, b, c) is a Pythagorean triple. Our task is to find integers u and v satisfying the four conditions. Let $u = c + b$ and $v = c - b$. Since b and c are positive and $c > b$, we have $u > v > 0$, so that (i) is satisfied. Also, $u - v = 2b$, so $2|u - v$ and condition (ii) holds. Further, $uv = (c + b)(c - b) = c^2 - b^2 = a^2$, so (iii) is satisfied. Finally, because $uv = a^2$, $a = \sqrt{uv}$, while $(u - v)/2 = (2b)/2 = b$ and $(u + v)/2 = 2c/2 = c$, showing that (iv) holds. This completes the proof. ∎

EXAMPLE. As an illustration of how to use this theorem, let us find all Pythagorean triples that have 30 as one of their legs. Certainly, if $(x, 30, z)$ is such a triple, then $(30, x, z)$ is also. Thus it is enough to find all Pythagorean triples (a, b, c) having $a = 30$. Now we want to find all pairs of integers u and v satisfying conditions (i), (ii), and (iii) of Theorem 7.2 and such that $a = \sqrt{uv} = 30$. Thus $uv = 30^2$, and uv is even. This means that at least one of u and v must be even. However, as we remarked in the proof, condition (ii) says that if one of u, v is even, so is the other. Thus both u and v must be even and $uv = 30^2$. Next, we claim that $u > 30$. For if $u \leq 30$, then by condition (i) we would have $30 \geq u > v > 0$, so $30^2 > uv = 30^2$. Thus our problem has been reduced to that of finding integers u and v such that

$$(1)\ u \mid 30^2; \quad (2)\ u > 30; \quad (3)\ u \text{ is even}; \quad (4)\ v = 30^2/u \text{ is even}$$

Because $30^2 = 2^2 \cdot 3^2 \cdot 5^2$, it is now an easy matter to find all such u. They are $u = 2 \cdot 3^2 \cdot 5^2 = 450$, $u = 2 \cdot 3^2 \cdot 5 = 90$, $u = 2 \cdot 3 \cdot 5^2 = 150$, and $u = 2 \cdot 5^2 = 50$. With $v = 30^2/u$ we get the following pairs (u, v):

$$(450, 2), \quad (90, 10), \quad (150, 6), \quad \text{and} \quad (50, 18)$$

Following the recipe in Theorem 7.2 and forming $a = \sqrt{uv}$, $b = (u - v)/2$ and $c = (u + v)/2$ with each of the pairs gives us the four Pythagorean triples

$$(30, 224, 226), \quad (30, 40, 50), \quad (30, 72, 78), \quad \text{and} \quad (30, 16, 34)$$

Therefore, these triples, together with their variations (224, 30, 226), and so on, are all the Pythagorean triples having 30 as a leg.

One might be interested in predicting the number of Pythagorean triples with a given leg a. The answer depends on whether a is odd or even. For a odd, it is $(\tau[a^2] - 1)/2$, while for even a it is $(\tau[(a/2)^2] - 1)/2$, with τ the function discussed in Chapter 5. (See Exercise 7.B.2.)

As we noted earlier, the treatment we have presented of Pythagorean triples is new, so we would like to briefly mention the older characterization. First note that Pythagorean triples fall into natural families such as:

$$(3, 4, 5), \quad (6, 8, 10), \quad (9, 12, 15), \quad (12, 16, 20), \ldots$$

or

$$(5, 12, 13), \quad (10, 24, 26), \quad (15, 36, 39), \quad (20, 48, 52), \ldots$$

That is, we have families of the form

$$(a, b, c), \quad (2a, 2b, 2c), \quad (3a, 3b, 3c), \quad (4a, 4b, 4c), \ldots$$

Note that $(3, 4) = 1$ and $(5,12) = 1$. This prompts a definition.

Definition. The Pythagorean triple (a, b, c) is said to be *primitive* if $(a, b) = 1$.

The next lemma is easy to prove.

Lemma 7.3. Suppose that (x, y, z) is a Pythagorean triple and let $d = (x, y)$. Then $d \mid z$. Further if a, b, and c are integers such that $x = ad$, $y = bd$, and $z = cd$, then (a, b, c) is a primitive Pythagorean triple.

What this lemma says then is that each Pythagorean triple (x, y, z) is the dth multiple of the primitive Pythagorean triple (a, b, c) obtained by dividing each of x, y, z by $d = (x, y)$. Thus, once we know all primitive Pythagorean triples, then we know all Pythagorean triples.

Lemma 7.4. If (a,b, c) is a primitive Pythagorean triple, then c is odd and $a \not\equiv b \bmod 2$.

Proof. If $a \equiv b \bmod 2$, then a and b are both even or both odd. If a and b are both even, then $2|(a, b)$, contrary to (a, b, c) being primitive. If a and b are both odd, then $c^2 = a^2 + b^2$ is even; hence c is even. Write $a = 2k + 1$, $b = 2n + 1$, and $c = 2m$. Then

$$\begin{aligned} a^2 &= 4k^2 + 4k + 1 \equiv 1 \bmod 4 \\ b^2 &= 4n^2 + 4n + 1 \equiv 1 \bmod 4 \\ c^2 &= 4m^2 \equiv 0 \bmod 4 \end{aligned}$$

so

$$a^2 + b^2 \equiv 2 \bmod 4$$

But $c^2 = a^2 + b^2$, so $0 \equiv 2 \bmod 4$, a contradiction. Thus $a \equiv b \bmod 2$. Since $a^2 + b^2 = c^2$ and exactly one of a or b is odd, c^2 must be odd and so c is odd. ▮

Theorem 7.5. Let a, b, and c be positive integers with a even. Then (a, b, c) is a primitive Pythagorean triple if and only if there exist integers x and y such that

1. $x > y > 0$.
2. $x \not\equiv y \bmod 2$.
3. $(x, y) = 1$.
4. $a = 2xy$, $b = x^2 - y^2$, and $c = x^2 + y^2$.

We will give an abbreviated proof of this theorem, leaving the details as an exercise (Exercise 7.A.8). The key idea is knowing how to find x and y, given a primitive Pythagorean triple (a, b, c). We illustrate with $(12, 5,13)$. Since 5 is odd, we consider $\dfrac{13 + 5}{12}$ and write this fraction in lowest terms; that is, $\dfrac{13 + 5}{12} = \dfrac{18}{12} = \dfrac{3}{2}$. We now take $x = 3$ and $y = 2$. Note that $2xy = 2 \cdot 3 \cdot 2 = 12 = a$, $x^2 - y^2 = 3^2 - 2^2 = 5 = b$, and $x^2 + y^2 = 3^2 + 2^2 = 13 = c$.

Proof (of Theorem 7.5). If such x and y exist, it is not difficult to see that $a = 2xy$, $b = x^2 - y^2$, and $c = x^2 + y^2$ give a primitive Pythagorean triple.

Thus assume that (a, b, c) is a primitive Pythagorean triple. By Lemma 7.4 we may assume that a is even. Suppose that $\frac{c+b}{a} = \frac{x}{y}$ with $\frac{x}{y}$ in lowest terms. It is not difficult to see that x and y satisfy (1) and (3). Now there is a $k \in Z$ such that $c + b = kx$ and $a = ky$. Furthermore, $\left(\frac{c+b}{a}\right)\left(\frac{c-b}{a}\right) = 1$, so that $\frac{c-b}{a} = \frac{y}{x}$ and there is an $l \in Z$ with $c - b = ly$ and $a = lx$. Since $lx = a = ky$ and $(x, y) = 1$, there is an $m \in Z$ with $l = my$ and $k = mx$. From $c + b = kx = mx^2$ and $c - b = ly = my^2$, by adding and subtracting we get $c = \frac{m(x^2 + y^2)}{2}$ and $b = \frac{m(x^2 - y^2)}{2}$. If $x \equiv y \bmod 2$, then $m \mid (c, b)$, so $m = 1$. As $a = ky = mxy$, we get x and y both even, so $2 \mid c$ and $2 \mid b$, a contradiction. Thus (2) holds. Also, $2 \mid m$ and $(m/2) \mid (c, b)$, so that $m/2 = 1$ and $m = 2$. Therefore, (4) holds. ∎

As a closing topic, let us illustrate how to find all *primitive* Pythagorean triples with a given leg a. We already know how to find all Pythagorean triples involving a. It requires finding all pairs of integers (u, v) satisfying (i), (ii), and (iii) of Theorem 7.2, with $a^2 = uv$, and then taking $(a, b, c) = \left(\sqrt{uv}, \frac{u-v}{2}, \frac{u+v}{2}\right)$. To get a primitive triple, we further insist that $\left(\sqrt{uv}, \frac{u-v}{2}\right) = 1$.

Let us first consider the case when a is odd; for example $a = 3^5 \cdot 5^2 \cdot 7 = 42525$. Since all the prime divisors of a are odd, by Exercise 7.A.9 to demand that $\left(\sqrt{uv}, \frac{u-v}{2}\right) = 1$ is the same as demanding that $(u, v) = 1$. Because $a^2 = 3^{10} \cdot 5^4 \cdot 7^2$, the pairs (u, v) satisfying our conditions are $(3^{10} \cdot 5^4 \cdot 7^2, 1)$, $(3^{10} \cdot 5^4, 7^2)$, $(3^{10} \cdot 7^2, 5^4)$, and $(3^{10}, 5^4 \cdot 7^2)$ (note that $3^{10} > a > 5^4 \cdot 7^2$). Using the recipe of Theorem 7.2, the primitive Pythagorean triples with $a = 42{,}525$ are (42,525, 904,187,812, 904,187,813), (42,525, 18,452,788, 18,452,837), (42,525, 1,446,388, 1,447,013), and (42,525, 14,212, 44,837).

The case when a is even is more involved. Let us take $a = 2^2 \cdot 3 \cdot 5 = 60$. As Theorem $7 \cdot 2$ requires that $u \equiv v \bmod 2$, we must now have u and v both even. Since $a^2 = 2^4 \cdot 3^2 \cdot 5^2$, Exercise 7.B.5 says that either $2 \mid u$ (and $2^2 \nmid u$) and $2^3 \mid v$, or $2 \mid v$ (and $2^2 \nmid v$) and $2^3 \mid u$. This, together with Exercise 7.A.9 lets us easily see that the possible (u, v) are $(2 \cdot 3^2 \cdot 5^2, 2^3)$, $(2^3 \cdot 3^2 \cdot 5^2, 2)$, $(2^3 \cdot 5^2,$

$2 \cdot 3^2$), and $(2^3 \cdot 3^2, 2 \cdot 5^2)$. The resulting primitive Pythagorean triples are (60, 221, 229), (60, 899, 901), (60, 91, 109), and (60, 11, 61).

Finally, note that Exercise 7.B.3 shows that if $2 \mid a$ but $4 \nmid a$, then there are no primitive Pythagorean triples having a as a leg. Otherwise, Exercise 7.B.6 tells how many primitive Pythagorean triples have a as their first leg.

7.2 Sums of Two Squares

Our objective in this section is to determine those integers n for which there exist integral solutions to the equation $X^2 + Y^2 = n$. Here we allow X or Y to be 0, as in $0^2 + 2^2 = 4$, so that 4 is one such n. Others are $n = 5$ and $n = 8$ since $2^2 + 1^2 = 5$ and $2^2 + 2^2 = 8$. It is easily seen that if $n = 6$ or $n = 7$, then $X^2 + Y^2 = n$ has no integral solutions.

Our first lemma was discovered by Euler (although it may well have been known to Diophantos). We say "discovered" rather than "proved" because its proof is trivial; any competent high school student could prove it, once told that it is true. Therefore, the realization that it is true represents the accomplishment.

Lemma 7.6. If $a, b, c, d \in Z$, then $(a^2 + b^2)(c^2 + d^2) = (ac + bd)^2 + (ad - bc)^2$. In particular, if each of n and m is a sum of two squares, then nm is a sum of two squares.

Our next result is the heart of the matter. It states that any prime that is congruent to 1 modulo 4 is a sum of two squares. Fermat proved this, noting the result in the margin of his copy of Diophantos. He did not give the proof. Some ninety years later, Euler set about supplying a proof; he succeeded after seven years of effort. Incidently, you may wonder how Fermat acquired his excellent reputation, when he so often withheld proofs. First, he did reveal more than enough of his work to establish himself as a genius of the first magnitude. Second, even the theorems he stated without proof were invariably later found to be true (with the possible exception of his Last Theorem). More important, his theorems invariably proved to be deep and important results which greatly influenced the work of later generations.

Theorem 7.7. If p is a prime integer and $p \equiv 1 \bmod 4$, then p can be written as the sum of two squares.

We give two different proofs.

First Proof. By Theorem 4.20 there is an integer t such that $t^2 + 1 \equiv 0 \bmod p$. Suppose that we can find integers a and b, not both equal to 0 and satisfying $a \equiv bt \bmod p$, $|a| < \sqrt{p}$ and $|b| < \sqrt{p}$. (We will soon see that we can do this.) Then

$$a^2 + b^2 \equiv (bt)^2 + b^2 = b^2(t^2 + 1) \equiv b^2 0 = 0 \bmod p$$

Thus $p \mid a^2 + b^2$ and we have $a^2 + b^2 = mp$ for some $m \in Z$. Since at least one of a and b is not zero, $a^2 + b^2 > 0$. Thus $m > 0$. Also,

$$mp = a^2 + b^2 = |a|^2 + |b|^2 < (\sqrt{p})^2 + (\sqrt{p})^2 = 2p$$

Thus $m < 2$. Since $0 < m < 2$, we must have $m = 1$, so $a^2 + b^2 = mp = p$.

Thus our proof hinges on being able to find integers a and b having the given properties. Let k be the largest integer such that $k^2 < p$. Consider the array

$$\begin{array}{ccccc} 0 - 0t & 0 - 1t & 0 - 2t & \cdots & 0 - kt \\ 1 - 0t & 1 - 1t & 1 - 2t & \cdots & 1 - kt \\ \vdots & & & & \\ k - 0t & k - 1t & k - 2t & \cdots & k - kt \end{array}$$

This array has $k + 1$ rows and $k + 1$ columns, and therefore contains $(k + 1)^2$ entries. As k is the largest integer satisfying $k^2 < p$, we must have $(k + 1)^2 > p$ (equality being impossible since p is prime). That is, there are more than p entries in the array. However, modulo p, there are only p different congruence classes of numbers, namely $0 \bmod p$, $1 \bmod p$, $2 \bmod p, \ldots, p - 1 \bmod p$. Thus two distinct entries in this array must fall in the same congruence class modulo p, and therefore are congruent to each other modulo p. Thus for some distinct pair of entries $a_1 - b_1 t$ and $a_2 - b_2 t$, we have $a_1 - b_1 t \equiv a_2 - b_2 t \bmod p$. Of course, $0 \le a_1 \le k, 0 \le b_1 \le k, 0 \le a_2 \le k, 0 \le b_2 \le k$, and since these two entries are distinct, either $a_1 \ne a_2$ or $b_1 \ne b_2$ (or both). Now let $a = a_1 - a_2$ and $b = b_1 - b_2$. We claim that these are the a and b desired in the first part of this proof. Because either $a_1 \ne a_2$ or $b_1 \ne b_2$, we have either $a \ne 0$ or $b \ne 0$. Because $0 \le a_1 \le k$ and $0 \le a_2 \le k$, clearly $-k \le a \le k$; that is, $|a| \le k$. However, $k^2 < p$, so that $k < \sqrt{p}$, so $|a| < \sqrt{p}$. In a similar fashion, $|b| < \sqrt{p}$. Finally, since $a_1 - b_1 t \equiv a_2 - b_2 t \bmod p$, we have

$$a - bt = (a_1 - a_2) - (b_1 - b_2)t = (a_1 - b_1 t) - (a_2 - b_2 t) \equiv 0 \bmod p$$

Thus $a \equiv bt \bmod p$, which completes the proof. ∎

The existence of the integers a and b which we have just used is also useful in other theorems in number theory. The proof that they exist (i.e., the last paragraph of the proof above) is due to the Norwegian mathematician Axel

Thue (1863–1922). The counting principle it used, showing that there are more entries in the table than there are congruence modulo p, is a special case of "Dirichlet's Box Principle," which is often also called the "Pigeon-hole Principle."

We wish to give a second proof of Theorem 7.7. The arguments in this second proof will appear again in the next section. First, we need a lemma.

Lemma 7.8. Let p be a prime integer with $p \equiv 1 \bmod 4$. Then there is an integer k with $1 \leq k < p$ such that

$$X^2 + 1 = kp$$

has a solution.

Proof. By Theorem 4.20 there is an integer t with $t^2 + 1 \equiv 0 \bmod p$. Of course, we may also assume that $0 \leq t \leq p - 1$. Now $t^2 + 1 \equiv 0 \bmod p$ means that $t^2 + 1 = kp$ for some $k \in Z$. However, $0 \leq t \leq p - 1$ implies that

$$kp = t^2 + 1 \leq (p - 1)^2 + 1 < p^2$$

Thus $k < p$. Also, $kp = t^2 + 1 > 0$ implies that $k > 0$. Therefore, $1 \leq k < p$ and letting $X = t$, we are done. ∎

We now give a second proof of Theorem 7.7.

Second Proof. Consider the set

$$S = \{m' \in Z \mid 1 \leq m' < p \text{ and } m'p \text{ is a sum of two squares}\}$$

We apply the principle of well-ordering to S. By Lemma 7.8, there is an integer k with $1 \leq k < p$ and kp is a sum of two squares (since $1 = 1^2$), so that $k \in S$. Thus $S \neq \varnothing$. Also if $m' \in S$, then $m' \geq 1$. Thus by well-ordering, S contains a least element, which we will call m. In particular, mp is a sum of two squares. The remainder of this proof will show that $m = 1$, so that $mp = p$ is a sum of two squares.

Now mp is a sum of two squares. Say

$$mp = a^2 + b^2 \tag{1}$$

Choose integers c and d satisfying the following conditions:

$$c \equiv a \bmod m \quad \text{and} \quad d \equiv b \bmod m \tag{2}$$

$$-m/2 < c \leq m/2 \quad \text{and} \quad -m/2 < d \leq -m/2 \tag{3}$$

[Note that there are m consecutive integers i which satisfy $-m/2 < i \leq m/2$, so that it is possible to find c and d satisfying (2) and (3).] From (1) and (2) we see that $c^2 + d^2 \equiv a^2 + b^2 = mp \equiv 0 \bmod m$. Thus $m|(c^2 + d^2)$, so we may write

$$(4) \qquad c^2 + d^2 = m^*m$$

for some $m^* \in \mathbb{Z}$. Now using (1), (4), and Lemma 7.6, we have $m^2m^*p = (mp)(m^*m) = (a^2 + b^2)(c^2 + d^2) = (ac + bd)^2 + (ad - bc)^2$. That is,

$$(5) \qquad m^2m^*p = (ac + bd)^2 + (ad - bc)^2$$

Using (1) and (2), we see that $ac + bd \equiv a^2 + b^2 = \text{mp} \equiv 0 \bmod m$, and $ad - bc \equiv ab - ba = 0 \bmod m$, which means that both $(ac + bd)/m$ and $(ad - bc)/m$ are integers. Therefore, we may divide both sides of (5) by m^2, arriving at

$$(6) \qquad m^*p = [(ac + bd)/m]^2 + [(ad - bc)/m]^2$$

That is, m^*p is the sum of two squares. In particular, we must have $m^* \geq 0$. Moreover, from (3) we have $c^2 = |c|^2 \leq m^2/4$ and $d^2 = |d|^2 \leq m^2/4$, so that by (4) we get

$$(*) \qquad m^*m = c^2 + d^2 \leq m^2/4 + m^2/4 = m^2/2 = (m/2)m$$

Dividing by m, we have $m^* \leq m/2 < m < p$. That is, $0 \leq m^* < m$ and m^*p is a sum of two squares. However, since m is the least element of S, and $m^* < m$, we have $m^* \notin S$. This shows that $m^* = 0$ (for if $m^* > 0$, we would have $1 \leq m^* < m < p$ and m^*p a sum of two squares, putting m^* in S, a contradiction). By (4), $c^2 + d^2 = m^*m = 0m = 0$. This can happen only if $c = 0$ and $d = 0$. Now by (2), $0 = c \equiv a \bmod m$ and $0 = d \equiv b \bmod m$, showing that $m|a$ and $m|b$. Therefore, $m^2|(a^2 + b^2)$, and by (1), $m^2|mp$. We therefore see that $m|p$. However, we already have $1 \leq m < p$ (since $m \in S$); because p is a prime and $m|p$, it must be that $m = 1$. This shows that $p = mp$ is a sum of two squares, completing the proof. ∎

Could either of these proofs be what Fermat had in mind? Apparently not, for in a letter to a friend he describes his proof as follows: "If an arbitrarily chosen prime of the form $4n + 1$ is not a sum of two squares . . . there will be another of the same nature, less than the one chosen . . . next a third still less, and so on. Making an infinite descent in this way we finally arrive at the number 5 . . . it follows that 5 is not the sum of two squares. But it is."

We now know that primes such as 13, 17, and 29 are each a sum of two squares. By Lemma 7.6, $13 \cdot 17$, $13 \cdot 29$, and $13 \cdot 17 \cdot 29$ are each a sum of two squares. Since $2 = 1^2 + 1^2$, Lemma 7.6 also tells us that $2 \cdot 13 \cdot 17$, $2 \cdot 13 \cdot 29$, and $2 \cdot 13 \cdot 17 \cdot 29$ are each a sum of two squares. For exercise, let us actually use Lemma 7.6 to find two squares whose sum is $6409 = 13 \cdot 17 \cdot 29$. Now $13 = 3^2 + 2^2$ and $17 = 4^2 + 1^2$, so by Lemma 7.6, $13 \cdot 17 = (3^2 + 2^2)(4^2 + 1^1)^2 = (3 \cdot 4 + 2 \cdot 1)^2 + (3 \cdot 1 - 2 \cdot 4)^2 = 14^2 + (-5)^2$. Now $29 = 5^2 + 2^2$, so $13 \cdot 17 \cdot 29 = [14^2 + (-5)^2][5^2 + 2^2] = [14 \cdot 5 + (-5)(2)]^2 + [14 \cdot 2 - (-5)(5)]^2 = 60^2 + 53^2$.

When you saw $13 \cdot 17 = 14^2 + (-5)^2$, hopefully you wondered what would happen if instead we wrote $13 \cdot 17 = 14^2 + 5^2$. Well, $13 \cdot 17 \cdot 29 = (14^2 + 5^2)(5^2 + 2^2) = (14 \cdot 5 + 5 \cdot 2)^2 + (14 \cdot 2 - 5 \cdot 5)^2 = 80^2 + 3^2$. Now we have two different ways of writing 6409 as a sum of two squares, and in general we can ask how many ways can a given n be expressed as a sum of two squares? (If any—and we have yet to completely answer that first question.) We will not treat this second question, except to say that the answer is known. In particular, if p is prime and if $p \equiv 1 \bmod 4$, then there is a unique way of expressing p as a sum of two squares. Thus $5^2 + 2^2 = 29$ is the only way of writing 29. [Of course, we are ignoring trivial variations such as $(-5)^2 + 2$ or $(-5)^2 + (-2)^2$.]

Returning now to the original question of which integers n are expressible as the sum of two squares, we see that $p = 2, 5, 13, 17, 29, 37, 41, \ldots$, and any products of these, are all sums of two squares. What about primes congruent to 3 modulo 4? You can quickly check that none of 3, 7, 11, 19, or 23 is a sum of two squares and you no doubt (correctly) guess that this is always the case. Furthermore, you can easily check that $3 \cdot 7 = 21$, $3 \cdot 11 = 33$, and $3 \cdot 7 \cdot 11 = 231$ are not sums of two squares. Of course, $7^2 = 7^2 + 0^2$, $3^4 = (3^2)^2 + 0^2$, and $3^2 \cdot 13 \cdot 17 \cdot 29 = 3^2(6409) = 3^2(60^2 + 53^2) = (3 \cdot 60)^2 + (3 \cdot 53)^2 = 180^2 + 159^2$, so that only odd powers of these primes cause concern. A little more experimentation and you will no doubt guess the general result. To state and prove it succinctly, we make the following definition.

Definition. The integer $m > 0$ is *square-free* if for each prime p, $p^2 \nmid m$.

For example, $70 = 2 \cdot 5 \cdot 7$ is square-free, as are $77 = 7 \cdot 11$ and $30 = 2 \cdot 3 \cdot 5$. However, $20 = 2^2 \cdot 5$ and $270 = 2 \cdot 3^3 \cdot 5$ are not square-free, since $2^2 \mid 20$ and $3^2 \mid 270$. Notice that 1 is square-free.

We use the following easy fact.

Lemma 7.9. Let n be a positive integer. Then there is a unique way of writing $n = k^2 m$ with k and m positive integers, and m square-free.

We leave the proof of this lemma as an exercise (7.A.24), but illustrate it with a few examples. $20 = 2^2 \cdot 5$ and 5 is square-free; $270 = 2 \cdot 3^3 \cdot 5 = 3^2(2 \cdot 3 \cdot 5)$ and $2 \cdot 3 \cdot 5$ is square-free; $567{,}000 = 2^3 \cdot 3^4 \cdot 5^3 \cdot 7 = 2^2 \cdot 3^4 \cdot 5^2 \cdot (2 \cdot 5 \cdot 7) = (2 \cdot 3^2 \cdot 5)^2(2 \cdot 5 \cdot 7)$ and $2 \cdot 5 \cdot 7$ is square-free; and $36 = 6^2 = 6^2 \cdot 1$

and 1 is square-free. In each case you can see that there is no other such representation.

Theorem 7.10. Let $n > 1$ be an integer with $n = k^2m$, where k and m are positive integers, m being square-free. Suppose that the prime decomposition of m is $m = p_1p_2\cdots p_t$. Then n is a sum of two squares if and only if $p_i \not\equiv 3 \bmod 4$ for each $i = 1, 2, \ldots, t$.

Proof. First notice that since m is square-free, we are correct in expressing $m = p_1p_2\cdots p_t$, with each exponent equal to 1.

Now suppose that $p_i \not\equiv 3 \bmod 4$ for each $i = 1, 2, \ldots, t$. Then since p_i is prime, either $p_i = 2$ or $p_i \equiv 1 \bmod 4$. By Theorem 7.7 (and since $2 = 1^2 + 1^2$), each p_i is a sum of two squares. Now by Lemma 7.6, $m = p_1p_2\cdots p_t$ is a sum of two squares, so we may write $m = a^2 + b^2$. Then $n = k^2m = k^2(a^2 + b^2) = (ka)^2 + (kb)^2$, so that n is a sum of two squares.

Conversely, suppose that $n = x^2 + y^2$. We must show that $p_i \not\equiv 3 \bmod 4$ for each $i = 1, 2, \ldots, t$. Let $d = (x, y)$ and suppose that $x = dx_0$, $y = dy_0$. Then $(x_0, y_0) = 1$ (Lemma 2.11). Let $n_0 = x_0^2 + y_0^2$, so that

$$n = x^2 + y^2 = (dx_0)^2 + (dy_0)^2 = d^2(x_0^2 + y_0^2) = d^2n_0$$

If we write $n_0 = k_0^2m_0$ with k_0 and m_0 positive integers and m_0 square-free, then

$$k^2m = n = d^2n_0 = d^2(k_0^2m_0) = (dk_0)^2m_0$$

Since m and m_0 are both square-free, the uniqueness statement in Lemma 7.9 shows that $m_0 = m = p_1p_2\cdots p_t$.

We now have $n_0 = x_0^2 + y_0^2$ with $(x_0, y_0) = 1$, $n_0 = k_0^2m_0$ and $m_0 = p_1p_2\cdots p_t$. Let us consider a fixed p_i, $i = 1, 2, \ldots, t$. Since $(x_0, y_0) = 1$, p_i cannot divide both of x_0 and y_0. Without harm we may assume that $p_i \nmid x_0$. Thus $(p_i, x_0) = 1$ and by Theorem 4.3 there is an integer z satisfying $zx_0 \equiv 1 \bmod p_i$. Now

$$z^2n_0 = z^2(x_0^2 + y_0^2) = (zx_0)^2 + (zy_0)^2 \equiv 1 + (zy_0)^2 \bmod p_i$$

However, $p_i \mid m_0$ and $m_0 \mid n_0$, so that $zn_0 \equiv 0 \bmod p_i$. Thus $1 + (zy_0)^2 \equiv 0 \bmod p_i$, so that zy_0 is a solution of $X^2 \equiv -1 \bmod p_i$. According to Theorem 4.20, $p_i \not\equiv 3 \bmod 4$ and our proof is complete. ∎

Notice the role that Theorem 4.20 played in the proof of Theorem 7.10. The second half of that proof used Theorem 4.20 directly. The first half of the proof used Theorem 7.7. We gave two proofs of Theorem 7.7, and both of them depended on Theorem 4.20. Thus Theorem 7.10 is seen to be a powerful generalization of the fact that for p a prime, $X^2 \equiv -1 \bmod p$ has a solution if and only if $p \not\equiv 3 \bmod 4$.

7.3 Sums of Four Squares

As the final topic of this chapter we will prove that each positive integer is a sum of four squares. Diophantos apparently guessed that this was true, although the earliest known record of the statement (made as a conjecture) is due to Bachet. No doubt you have surmised who first proved the theorem —Fermat. Again the margin of his copy of Diophantos held the theorem but not the proof. Again Euler took up the gauntlet. This time, however, Euler needed help. He did come very close, proving Lemmas 7.11 and 7.12. Yet it fell to the French mathematician Lagrange to incorporate Euler's work into the first published proof. A few years later, Euler found an easier proof of his own. This was some 40 years after he began working on the problem!

Our first lemma is the analogue of Lemma 7.6. This one was harder to discover. It took Euler 13 years to find it.

Lemma 7.11. If $a_1, b_1, c_1, d_1, a_2, b_2, c_2, d_2 \in Z$, then $(a_1^2 + b_1^2 + c_1^2 + d_1^2)\cdot(a_2^2 + b_2^2 + c_2^2 + d_2^2) = e_1^2 + e_2^2 + e_3^2 + e_4^2$, where

$$\begin{aligned}
e_1 &= a_1a_2 + b_1b_2 + c_1c_2 + d_1d_2 \\
e_2 &= a_1b_2 - b_1a_2 + c_1d_2 - d_1c_2 \\
e_3 &= a_1c_2 - c_1a_2 + d_1b_2 - b_1d_2 \\
e_4 &= a_1d_2 - d_1a_2 + b_1c_2 - c_1b_2
\end{aligned}$$

In particular, if n and m are each the sum of four squares, then nm is the sum of four squares.

Proof. We leave the calculation to you, suggesting only that you first show that

$$(x + y + z + w)^2 = x^2 + y^2 + z^2 + w^2 + 2xy + 2xz + 2xw + 2yz + 2yw + 2zw$$

In view of Lemma 7.11, it is enough to show that each prime is a sum of four squares. Incidentally, the modern subject called quadratic forms continues the study of identities such as those in Lemmas 7.6 and 7.11.

Lemma 7.12. If $p > 2$ is a prime integer, then there is an integer k with $1 \leq k < p$ such that

$$X^2 + Y^2 + 1 = kp$$

has a solution.

Proof. Since $p > 2$, p is odd and $(p - 1)/2 \in \mathbb{Z}$. Consider the set

$$S = \left\{0^2, 1^2, 2^2, \ldots, \left(\frac{p-1}{2}\right)^2\right\}$$

Counting the 0^2, we see that this set contains $(p - 1)/2 + 1 = (p + 1)/2$ elements. We claim that they are all distinct modulo p. If not, suppose that $i^2 \equiv j^2 \bmod p$ with $0 \le i \le (p - 1)/2$, $0 \le j \le (p - 1)/2$ and $i \ne j$. Then i and j are both solutions of $X^2 \equiv i^2 \bmod p$. This contradicts Exercise 4.B.16(b).

We next look at

$$T = \{p - 1 - 0^2, p - 1 - 1^2, p - 1 - 2^2, \ldots, p - 1 - ((p - 1)/2)^2\}$$

This set also contains $(p + 1)/2$ elements and they must be distinct modulo p, for if $p - 1 - i^2 \equiv p - 1 - j^2 \bmod p$, Then obviously $i^2 \equiv j^2 \bmod p$, which we just saw cannot happen with $0 \le i \le p - 1/2$, $0 \le j \le (p - 1)/2$ unless $i = j$.

We now claim that there is an element a^2 in S and an element $p - 1 - b^2$ in T with

(i) $$a^2 \equiv p - 1 - b^2 \bmod p$$

If there were no such pair, that is, if each element of S was distinct modulo p from each element of T, then between them, S and T would contain $(p + 1)/2 + (p + 1)/2 = p + 1$ distinct elements modulo p. However, there are only p different congruence classes of integers modulo p, so that it is impossible to have $p + 1$ distinct elements modulo p. This proves that condition (i) holds as stated. Now from (i) we get $a^2 + b^2 + 1 \equiv 0 \bmod p$. Choose integers a_0 and b_0 satisfying

$$a_0 \equiv a \bmod p \qquad b_0 \equiv b \bmod p$$

and

$$-p/2 < a_0 \le p/2 \qquad -p/2 < b_0 \le p/2$$

Thus $|a_0| \le p/2$ and $|b_0| \le p/2$ and $a_0^2 + b_0^2 + 1 \equiv 0 \bmod p$. If we write $a_0^2 + b_0^2 + 1 = kp$, then

$$kp = a_0^2 + b_0^2 + 1 = |a_0|^2 + |b_0|^2 + 1 \le \left(\frac{p}{2}\right)^2 + \left(\frac{p}{2}\right)^2 + 1 = \frac{p^2}{2} + 1$$

Thus $2kp \le p^2 + 2 < 2p^2$, so that $k < p$. Also, since $a_0^2 + b_0^2 + 1 > 0$, $k > 0$. Thus $1 \le k < p$ and a_0, b_0 is a solution of $X^2 + Y^2 + 1 = kp$, proving the lemma. ∎

Lemma 7.13. Let p be a prime integer. Suppose that k is an integer and that

$$X^2 + Y^2 + Z^2 + W^2 = kp$$

has a solution. If k is even, then

$$X^2 + Y^2 + Z^2 + W^2 = (k/2)p$$

has a solution.

Proof. Suppose that $a^2 + b^2 + c^2 + d^2 = kp$. Since modulo 2 there are only two congruence classes of integers, some pair of elements among a, b, c, d must be congruent modulo 2. We may assume that $a \equiv b \bmod 2$. Now $a^2 \equiv b^2 \bmod 2$, so that $a^2 + b^2 \equiv 0 \bmod 2$. Since k is even, $kp \equiv 0 \bmod 2$, and from our equation we see that $c^2 + d^2 \equiv 0 \bmod 2$. Thus $c^2 \equiv d^2 \bmod 2$, so $c \equiv d \bmod 2$. We now have both $a \equiv b \bmod 2$ and $c \equiv d \bmod 2$. Thus $(a + b)/2$, $(a - b)/2$, $(c + d)/2$, and $(c - d)/2$ are all integers. A simple calculation shows that

$$\left(\frac{a+b}{2}\right)^2 + \left(\frac{a-b}{2}\right)^2 + \left(\frac{c+d}{2}\right)^2 + \left(\frac{c-d}{2}\right)^2$$
$$= (a^2 + b^2)/2 + (c^2 + d^2)/2 = kp/2 = (k/2)p$$

That is, $X^2 + Y^2 + Z^2 + W^2 = (k/2)p$ has a solution, as claimed. ∎

We now reach the crux of the matter, in a proof very similar to the second proof of Theorem 7.7.

Theorem 7.14. Every positive integer is a sum of four squares.

Proof. By Lemma 7.11 it is enough to show that each prime integer is a sum of four squares. Now $2 = 1^2 + 1^2 + 0^2 + 0^2$, so we consider a prime $p > 2$. Let

$$S = \{m' \in Z \mid 1 \leq m' < p \text{ and } m'p \text{ is a sum of four squares}\}$$

By Lemma 7.12, there is an integer k with $1 \leq k < p$ such that kp is a sum of three squares. Thus $k \in S$, so $S \neq \varnothing$. Also, if $m' \in S$, then $m' \geq 1$. Thus by the Principal of Well-Ordering, S contains a least element, which we will call m.

Now m must be odd, since if m is even, then Lemma 7.13 shows that $m/2 \in S$. This cannot be the case since m is the least element of S.

As $m \in S$, mp is a sum of four squares. Say that

$$(1') \qquad mp = a_1^2 + b_1^2 + c_1^2 + d_1^2$$

Now choose a_2, b_2, c_2, d_2 satisfying

$$(2') \qquad \begin{aligned} a_2 &\equiv a_1 \bmod m; & b_2 &\equiv b_1 \bmod m; \\ c_2 &\equiv c_1 \bmod m; & d_2 &\equiv d_1 \bmod m \end{aligned}$$

and

$$(3') \qquad \begin{aligned} -m/2 &< a_2 < m/2; & -m/2 &< b_2 < m/2; \\ -m/2 &< c_2 < m/2; & -m/2 &< d_2 < m/2 \end{aligned}$$

We can do this, since there are m consecutive integers i satisfying $-m/2 < i < m/2$. This is because m is odd, so that $m/2$ is not an integer.

From (1′) and (2′), $a_2^2 + b_2^2 + c_2^2 + d_2^2 \equiv a_1^2 + b_1^2 + c_1^2 + d_1^2 = mp \equiv 0 \bmod m$, so that

$$(4') \qquad a_2^2 + b_2^2 + c_2^2 + d_2^2 = m^*m$$

for some $m^* \in Z$. Now using (1′) and (4′),

$$m^2m^*p = (mp)(m^*m) = (a_1^2 + b_1^2 + c_1^2 + d_1^2)(a_2^2 + b_2^2 + c_2^2 + d_2^2)$$

so that by Lemma 7.11,

$$(5') \qquad m^2m^*p = e_1^2 + e_2^2 + e_3^2 + e_4^2$$

where

$$\begin{aligned} e_1 &= a_1a_2 + b_1b_2 + c_1c_2 + d_1d_2 \\ e_2 &= a_1b_2 - b_1a_2 + c_1d_2 - d_1c_2 \\ e_3 &= a_1c_2 - c_1a_2 + d_1b_2 - b_1d_2 \\ e_4 &= a_1d_2 - d_1a_2 + b_1c_2 - c_1b_2 \end{aligned}$$

However, we claim that $e_i \equiv 0 \bmod m$ for $i = 1, 2, 3, 4$. For e_1, by (2′) and (1′) we have $e_1 \equiv a_1^2 + b_1^2 + c_1^2 + d_1^2 \equiv mp \equiv 0 \bmod m$. For e_2, by (2′) we have $e_2 \equiv a_1b_1 - b_1a_1 + c_1d_1 - d_1c_1 = 0 \bmod m$, and e_3 and e_4 behave similarly. Thus m divides each of e_1, e_2, e_3, and e_4. Now dividing (5′) by m^2, we obtain

$$(6') \qquad m^*p = (e_1/m)^2 + (e_2/m)^2 + (e_3/m)^2 + (e_4/m)^2$$

Obviously, $m^* \geq 0$ and from (4′) and (3′),

$$(**) \quad m^*m = a_2^2 + b_2^2 + c_2^2 + d_2^2 < (m/2)^2 + (m/2)^2 + (m/2)^2 + (m/2)^2 = m^2$$

Dividing by m, $m^* < m$. That is, $0 \leq m^* < m$. Since m^*p is a sum of four squares, we must have $m^* = 0$, since otherwise $1 \leq m^* < m < p$ would imply that $m^* \in S$, contradicting that m is the least element of S. Now $m^* = 0$ together with (4′) shows that $a_2 = b_2 = c_2 = d_2 = 0$. From (2′), we now get that $a_1 \equiv b_1 \equiv c_1 \equiv d_1 \equiv 0 \bmod m$. Thus $m^2 | a_1^2 + b_1^2 + c_1^2 + d_1^2$, and by (1′), $m^2 | mp$. Thus $m | p$. However, $1 \leq m < p$, so $m = 1$. This shows that $p = mp$ is a sum of four squares, completing the proof. ∎

We would like to point out the similarities between the arguments involved in dealing with sums of two squares and sums of four squares. Of course, Lemma 7.6 and Lemma 7.11 are analogous. Similarly, Lemmas 7.8 and 7.12 are analogous. The (second) proof of Theorem 7.7 and the proof of Theorem 7.14 are almost identical. However, Lemma 7.13 has no counterpart in the arguments concerning sums of two squares. Such a counterpart exists (Exercise 7.A.29) but is simply not needed in the proof. We leave it as an exercise to explain this difference in the two sections (Exercise 7.B.12).

You may wonder about sums of three squares. We only mention that it is known that the positive integer n is a sum of three squares if and only if n is *not* of the form $n = 4^k m$ with $m \equiv 7 \bmod 8$.

Knowing that each positive integer is a sum of four squares, in 1770 Edward Waring conjectured that each positive integer was a sum of nine cubes (of positive integers) and a sum of 19 fourth powers (and that no fewer are enough). For instance, $23 = 2^3 + 2^3 + 1^3 + 1^3 + 1^3 + 1^3 + 1^3 + 1^3 + 1^3$ cannot be written as a sum of less than nine cubes, and 79 cannot be written as a sum of less than 19 fourth powers. [In dealing with cubes, we must omit negative integers, otherwise, five suffices: $x = x^3 + (k+1)^3 + (k-1)^3 + (-k)^3 + (-k)^3$ with $k = \dfrac{x - x^3}{6}$.] Furthermore, Waring conjectured for kth powers, there was a number $N(k)$ such that each positive integer could be written as $N(k)$ kth powers [and that some positive integer required $N(k)$ kth powers]. However, he did not venture a guess as to what $N(k)$ might be except for the cases $N(3) = 9$ and $N(4) = 19$. [Of course, $N(2) = 4$ was already known.] Later work showed that $N(3)$ did exist, and did equal 9. It was also shown that $N(4)$ existed but it has not yet been determined if it does equal 19. The German, David Hilbert, generally recognized as the finest mathematician of this century, proved that $N(k)$ does exist for each integer $k > 1$. However, his arguments gave no method of calculating what value a given $N(k)$ might have. There is a conjecture that $N(k) = 2^k + [(\frac{3}{2})^k] - 2$. Here the square bracket represents the "greatest integer" function. For example, $\left[2\frac{1}{2}\right] = 2$, $\left[3\frac{1}{10}\right] = 3$, $\left[17\frac{5}{23}\right] = 17$, $[6.007] = 6$. Thus this conjecture says that $N(2) = 4$, $N(3) = 9$, $N(4) = 19$, $N(5) = 37$, $N(6) = 73$, ... Considerable effort has gone into attempts to prove this conjecture and the only remaining cases of interest are $N(4)$ and $N(5)$.

As a footnote, we mention that the Wilson of Wilson's Theorem (Theorem 4.19) was John Wilson, and that during his brief career as a mathematician (he gave it up to study law) he was a student of Edward Waring.

EXERCISES 7.A

1. Verify that if n is a positive integer, then $(2n + 1, 2n^2 + 2n, 2n^2 + 2n + 1)$ is a Pythagorean triple.

2. Verify that if a is even, then $(a, (a^2 - 4)/4, (a^2 + 4)/4)$ is a Pythagorean triple.

3. Verify that if a is odd, $a > 2$, then $(a, (a^2 - 1)/2, (a^2 + 1)/2)$ is a Pythagorean triple.

4. For each of the following Pythagorean triples, find integers u and v that satisfy the conditions of Theorem 7.2.

(a) (6, 8, 10) (b) (16, 30, 34)
(c) (14, 48, 50) (d) (16, 63, 65)
(e) (4961, 6480, 8161)

5. For each of the following choices of a, find all Pythagorean triples having a as their first leg.

(a) 5 (b) 8
(c) 33 (d) 72

6. Show that if (a, b, c) is a Pythagorean triple and k is any positive integer then (ka, kb, kc) is a Pythagorean triple.

7. Prove Lemma 7.3.

8. (a) Read the outline of the proof of Theorem 7.5 compiling a list of statements that might require more detailed verification.

(b) Supply the needed verifications for your statements in (a).

9. Let $u > v > 0$ be integers such that uv is a perfect square and $u \equiv v \bmod 2$. If p is an odd prime, show that $p \mid (\sqrt{uv}, (u - v)/2)$ if and only if $p \mid (u, v)$.

10. Find all primitive Pythagorean triples having each of the following choices of a as the first leg.

(a) 5 (b) 15 (c) 25
(d) 105 (e) 6 (f) 12
(g) 72

11. Are there any Pythagorean triples (a, b, c) with $a = b$?

12. Are there any Pythagorean triples (a, b, c) with $a = 1$?

13. Are there any Pythagorean triples (a, b, c) with $a = 2$?

14. Show that if k is a positive integer, then $\sqrt{1 + k^2}$ is never an integer.

15. Are there positive integers a and k such that (a, ka, c) is a Pythagorean triple?

16. (a) Determine all solutions of the congruence $a^2 + b^2 \equiv c^2 \bmod 3$. [*Hint:* One solution is (0 mod 3, 1 mod 3, 2 mod 3).]

(b) Show that if (a, b, c) is a Pythagorean triple, then one of a or b is divisible by 3.

17. (a) Determine all solutions of the congruence $a^2 + b^2 \equiv c^2 \bmod 5$.

(b) Show that if (a, b, c) is a Pythagorean triple, then one of a, b, or c is divisible by 5.

18. (a) Determine all solutions of the congruence $a^2 + b^2 \equiv c^2 \bmod 7$.

(b) Show that if (a, b, c) is a Pythagorean triple, then one of a, b, $a + b$, or $a - b$ is divisible by 7.

19. Show that if (a, b, c) is a Pythagorean triple then $\dfrac{c+b}{a} = \sqrt{\dfrac{c+b}{c-b}}$.

20. If (a, b, c) is a primitive Pythagorean triple with a even, show that $c + a$ and $c - a$ are both squares while $c + b$ and $c - b$ are two times a square.

21. Do the computation necessary to prove Lemma 7.6.

22. Use Lemma 7.6 to write $1105 = 5 \cdot 13 \cdot 17$ as the sum of two squares. Do the same for $2465 = 5 \cdot 17 \cdot 29$ and $1885 = 5 \cdot 13 \cdot 29$.

23. Verify that Lemma 7.8 is true for each prime less than 20 by finding some x and k that make it true.

24. Prove Lemma 7.9.

25. Use Theorem 7.10 to predict which integers n between 40 and 60 can be written as the sum of two squares. Then for each such n, find two squares having n as their sum.

26. Do the computations necessary to prove Lemma 7.11.

27. Verify that Lemma 7.12 holds for each prime less than 20, by finding some X, Y, and k that make it true.

28. Starting with $3^2 + 4^2 + 5^2 + 6^2 = 86$, use the construction in the proof of Lemma 7.13 to find a solution of $X^2 + Y^2 + Z^2 + W^2 = 43$.

29. Let p be a prime integer. Suppose that k is an integer such that $X^2 + Y^2 = kp$ has a solution. Also suppose that k is even.

(a) Prove that $X^2 + Y^2 = \left(\dfrac{k}{2}\right)p$ has a solution, by mimicking the proof of Lemma 7.13.

(b) Give a second proof that $X^2 + Y^2 = \left(\dfrac{k}{2}\right)p$ has a solution, this time by using Theorem 7.10.

30. (a) For which of the following values of c can you find a Pythagorean triple (a, b, c)?

$$c = 3;\ c = 5;\ c = 7;\ c = 6;\ c = 10;\ c = 14;\ c = 9;\ c = 15$$

(b) Try to guess for what values of c (a) can be done.

31. In the first section of this chapter we mention a problem from the Palatine Anthology concerning Diophantos. Use it to calculate Diophantos' alleged age. (*Hint:* Interpret "half of his father he was" to mean "he lived half as long as his father lived.")

EXERCISES 7.B

1. Show that if a is odd, then there is a Pythagorean triple (a, b, c) such that $c = b + 1$.

2. Show that the number of Pythagorean triples (a, b, c) for a given integer a is given by

$$(\tau(a^2) - 1)/2 \qquad \text{if } a \text{ is odd}$$

$$\left(\tau\left(\left(\frac{a}{2}\right)^2\right) - 1\right)/2 \qquad \text{if } a \text{ is even}$$

3. Suppose that (a, b, c) is a primitive Pythagorean triple with b odd. Show that $4 \mid a$.

4. Show that if (a, b, c) is a Pythagorean triple such that $2 \mid a$ but $4 \nmid a$ then $8 \mid b$. (*Hint:* Use Exercise 7.B.3.)

5. Let m and n be odd integers such that mn is a perfect square. Suppose also that e is an integer, $e \geq 2$, and let $u = 2^i m$, $v = 2^{2e-1}n$, where $1 \leq i < 2e$.
(a) Show that if $i = 1$ or if $i = 2e - 1$, then $2 \nmid (\sqrt{uv}, (u - v)/2)$.
(b) Show that if $1 < i < 2e - 1$, then $2 \mid (\sqrt{uv}, (u - v)/2)$.

6. Let a be a positive integer with prime decomposition $a = p_1^{e_1} p_2^{e_2} \cdots p_n^{e_n}$. Show that the number of primitive Pythagorean triples with a as one leg is 2^{n-1} except in the case where $2 \mid a$ but $4 \nmid a$. (*Remark:* In case $2 \mid a$ and $4 \nmid a$, see Exercise 7.B.3.)

7. Let (a, b, c) be a primitive Pythagorean triple. Let p be prime with $p \mid c$. Show that $p \equiv 1 \bmod 4$. [*Hint:* By Lemma 7.4, $p \neq 2$. Now $(p, a) = 1$. Suppose that $da \equiv 1 \bmod p$. Then $0 \equiv 1 + (db)^2 \bmod p$. Use Theorem 4.20.]

8. Let $c > 0$. Show that there is a Pythagorean triple (a, b, c) if and only if c is divisible by some prime p with $p \equiv 1 \bmod 4$. [*Hint:* Use Exercise 7.B.7 for one direction. Conversely, if $p \mid c$, $p \equiv 1 \bmod 4$, write $p = x^2 + y^2$. Note that $x \neq y$. Assume that $x > y$. Consider $(2xy, x^2 - y^2, x^2 + y^2)$.]

9. If you are not already familiar with it, read about the "norm" function for complex numbers (discussed in most calculus texts). Explain what Lemma 7.6 tells you about it.

10. In the second proof of Theorem 7.7, we have marked one equation by (*). In the proof of Theorem 7.14, we have marked one equation by (**). Compare these two equations and explain why in the second one we must have strict inequality (i.e., " $<$ ") while in the first one we are content with just " $\leq$." (*Hint:* $m/2 < m$ but $m \leq m$.)

11. Consider the equation marked (**) in the proof of Theorem 7.14. Explain how the fact that m was odd allowed us to have " $<$ " instead of just " $\leq$ " in this equation.

12. Explain the role that Lemma 7.13 plays in the proof of Theorem 7.14, and why the second proof of Theorem 7.7 did not require an analogous result.

From Diophantos' *Arithmetics**

PROBLEM 29, BOOK II: Find two square numbers such that when one forms their product and adds either of the numbers to it, the result is a square.

PROBLEM 7, BOOK III: Find three numbers such that their sum is a square and the sum of any two of them is a square.

PROBLEM 9, BOOK III: Find three numbers in arithmetic series such that the sum of any two of them is a square.

PROBLEM 15, BOOK III: Find three numbers such that the product of two of them minus the third is always a square.

PROBLEM 18, BOOK VI: Find a Pythagorean triangle in which the length of the bisector of one of the acute angles is rational.

* From *Number Theory and Its History* by Oystein Ore. Copyright 1948 by McGraw-Hill Book Company, Inc. Used by permission of McGraw-Hill Book Company.

8

Primitive Roots

IN Chapter 4 we saw that if m is a positive integer and $(b, m) = 1$, then the order of b modulo m is a divisor of $\varphi(m)$. We also asked for what m is the converse true? That is, for what positive integers m is it true that every positive divisor of $\varphi(m)$ is in fact the order of some integer b modulo m [with $(b, m) = 1$, as always]? In this chapter we will answer that question and we will show you a few of the nice things that happen for those m. The next definition is fundamental.

Definition. Suppose that m is a positive integer and $a \in \mathbb{Z}$ with $(a, m) = 1$. We say that a is a *primitive root of* m if the order of a modulo m is $\varphi(m)$.

Here is an example of what we mean. If $m = 11$, you can easily check that each of the integers 2, 6, 7, 8 has order $10 = \varphi(11)$ modulo 11. Thus these four integers are each primitive roots of 11. In contrast, for $m = 8$ there are no primitive roots, since $\varphi(8) = 4$ and none of 1, 3, 5, and 7 have order 4 modulo 8, which again is easily checked.

Theorem 8.1. Suppose that the positive integer m has a primitive root. If d is a positive integer and $d \mid \varphi(m)$, then there is an integer b such that $(b, m) = 1$ and the order of b modulo m is d.

Proof. Let a be a primitive root of m. Since $d \mid \varphi(m)$, we have $\varphi(m) = d \cdot k$, where $k > 0$. Our choice for b is $b = a^k$. By the definition of primitive root $(a, m) = 1$ and hence $(b, m) = 1$. We must now show that the order of b modulo m is d. Now

$$b^d = (a^k)^d = a^{\varphi(m)} \equiv 1 \bmod m$$

Thus the order of b modulo m must divide d. Let us denote the order of b modulo m by d_1. We have $d_1 \mid d$, so that $d_1 \leq d$. Hence $d_1 k \leq dk = \varphi(m)$. Since the order of b modulo m is d_1, we have

$$a^{kd_1} = (a^k)^{d_1} = (b)^{d_1} \equiv 1 \bmod m$$

Because a is a primitive root of m, the order of a modulo m is $\varphi(m)$. By the definition of order we must have $\varphi(m) \leq kd_1$. We already have $kd_1 \leq kd = \varphi(m)$; thus $kd_1 = kd$, so $d_1 = d$; that is, the order of b modulo m is d, as desired. ∎

Recalling the question raised at the beginning of this chapter, we now see that the m which answer that question are exactly the m that have primitive roots. We have just seen that an m possessing a primitive root answers our question. Conversely, if $m > 0$ has the property that for each positive divisor d of $\varphi(m)$, there is an integer whose order modulo m is d, then letting $d = \varphi(m)$ we see that m has a primitive root.

Proposition 8.2. Let $m > 0$ and assume that a is a primitive root of m. If $(b, m) = 1$, then there is an integer $k \geq 1$ such that $b \equiv a^k \bmod m$.

Proof. Because the order of a is $\varphi(m)$, the powers $a, a^2, \ldots, a^{\varphi(m)}$ are all distinct modulo m (Theorem 4.7) and $(a^k, m) = 1$ for all $k \geq 1$. The set $\{a, a^2, \ldots, a^{\varphi(m)}\}$ contains $\varphi(m)$ distinct integers modulo m, and modulo m there are exactly $\varphi(m)$ integers relatively prime to m. If $(b, m) = 1$, then b is congruent to some power of a modulo m. ∎

Let us look at an example to see how Proposition 8.2 can be put to use.

EXAMPLES

(a) Choose $m = 14$. The positive integers less than 14 and relatively prime to 14 are 1, 3, 5, 9, 11, and 13. Thus $\varphi(14) = 6$. Now modulo 14,

$$3^1 \equiv 3, \quad 3^2 \equiv 9, \quad 3^3 \equiv 13, \quad 3^4 \equiv 11, \quad 3^5 \equiv 5, \quad \text{and} \quad 3^6 \equiv 1$$

Therefore, the order of 3 modulo 14 is $6 = \varphi(14)$, so 3 is a primitive root of 14.

(b) If we wish to solve a congruence such as

$$9X \equiv 5 \bmod 14$$

we might argue that $9 \equiv 3^2 \bmod 14$ and $5 \equiv 3^5 \bmod 14$, so that

$$3^2X \equiv 3^5 \bmod 14$$

Because $(3^2, 14) = 1$, we may cancel and arrive at

$$X \equiv 3^3 \equiv 13 \bmod 14$$

as the solution.

(c) If we were faced with the congruence

$$X^4 \equiv 9 \bmod 14$$

we would say that $9 \equiv 3^2 \bmod 14$ and so we want

$$X^4 \equiv 3^2 \bmod 14$$

Since $(3, 14) = 1$, we must have $(X, 14) = 1$ and X must be congruent to some positive power of 3 modulo 14. Let us suppose that $X \equiv 3^k \bmod 14$. Then

$$3^{4k} \equiv 3^2 \bmod 14$$

and so canceling 3^2 we arrive at

$$3^{4k-2} \equiv 1 \bmod 14$$

Now 3 has order $\varphi(14) = 6$, so by Theorem 4.6, $6 \mid 4k - 2$. Put another way, $4k \equiv 2 \bmod 6$. From Chapter 3 we know all about solving this congruence. It has two solutions modulo 6:

$$k \equiv 2, \qquad k \equiv 5 \bmod 6$$

Thus substituting these two values in $X \equiv 3^k \bmod 6$, we obtain the two solutions

$$X \equiv 3^2 \equiv 9 \qquad \text{and} \qquad X \equiv 3^5 \equiv 5 \bmod 14$$

(d) Had you started with the congruence

$$X^4 \equiv 13 \bmod 14$$

then, since $13 \equiv 3^3$, you would wind up having to solve the congruence

$$4k \equiv 3 \bmod 6$$

As you can see, there are no solutions to this congruence, so there are no solutions to the congruence $X^4 \equiv 13 \bmod 14$.

These examples should make it clear that if a positive integer m has a primitive root, then it is easy to solve equations of the form $bX \equiv c \bmod m$, or of the form $X^n \equiv c \bmod m$ for b and c relatively prime to m. There are several lovely theorems which can be proved along these lines. We leave these as exercises (Exercises 8.B.1 through 3), however. We will only prove one very easy result, because it will quickly help us.

Theorem 8.3. If $m > 2$ has a primitive root, then the congruence

$$X^2 \equiv 1 \bmod m$$

has exactly two solutions, namely $X \equiv 1 \bmod m$ and $X \equiv -1 \bmod m$.

Proof. Notice that $X \equiv 1$ and $X \equiv -1$ are both solutions of the congruence, so what we must establish is that any solution is congruent modulo m to 1 or -1. Let a be a primitive root of m. Suppose that c is a solution to the congruence $X^2 \equiv 1 \bmod m$. Then $c^2 \equiv 1 \bmod m$, so that $(c, m) = 1$. By Proposition 8.2, there is an integer $k \geq 1$ such that $a^k \equiv c \bmod m$. Then

$$a^{2k} \equiv c^2 \equiv 1 \bmod m$$

Because a has order $\varphi(m)$ we must have $\varphi(m) | 2k$. Thus k is a solution of the congruence

$$2Y \equiv 0 \bmod \varphi(m)$$

By Theorem 6.2, there are exactly $(2, \varphi(m))$ solutions modulo $\varphi(m)$ to this congruence. Because $m > 2$, $\varphi(m)$ is even (Exercise 5.A.13) and $(2, \varphi(m)) = 2$. Therefore, there are exactly two solutions to the congruence $X^2 \equiv 1 \bmod m$. Since 1 and -1 are both solutions, we conclude that $c \equiv 1$ or $c \equiv -1 \bmod m$. ∎

Obviously, Theorem 8.3 is useful. Consider the congruence

$$X^2 \equiv 1 \bmod 15$$

Since $4^2 \equiv 1 \bmod 15$ and $4 \not\equiv \pm 1 \bmod 15$, Theorem 8.3 tells us that 15 does *not* have a primitive root. Using this idea more formally, we have the following.

Theorem 8.4. Let m be an integer that has a primitive root. Then m has the form 2, 4, p^n or $2p^n$ with p an odd prime and $n \geq 1$.

Proof. Suppose first that it is possible to write m as $m = m_1 m_2$, where $(m_1, m_2) = 1$ and $m_1 > 2$, $m_2 > 2$. We claim that under these circumstances m cannot have a primitive root. For consider the congruence $X^2 \equiv 1 \bmod m_1$. This congruence has at least two distinct solutions $X \equiv 1 \bmod m_1$ and $X \equiv -1 \bmod m_1$ ($1 \not\equiv -1 \bmod m_1$ because $m_1 > 2$). Similarly, $X^2 \equiv 1 \bmod m_2$ has at least two distinct solutions, $X \equiv 1 \bmod m_2$ and $X \equiv -1 \bmod m_2$. Then by Corollary 6.13, $X^2 \equiv 1 \bmod m_1 m_2$ has at least four distinct solutions. Accordingly, m cannot have a primitive root.

The preceding paragraph tells us that if m has a primitive root, then m cannot be expressed as a product of two relatively prime integers, each greater than 2. From prime decomposition it now follows easily that if m has a primitive root, then m has the form $m = p^n$ with p a prime or $m = 2 \cdot p^n$ with p an odd prime.

To complete the proof, it only remains to show that for $n \geq 3$, $m = 2^n$ does not have a primitive root. Again we use Theorem 8.3 and examine the congruence $X^2 \equiv 1 \bmod 2^n$. We have

$$(2^{n-1} - 1)^2 = 2^{2n-2} - 2^n + 1 = 2^n(2^{n-2} - 1) + 1 \equiv 1 \bmod 2^n$$

Furthermore, $2^{n-1} - 1 \not\equiv \pm 1 \bmod 2^n$, so there are at least three distinct solutions to $X^2 \equiv 1 \bmod 2^n$. Therefore, 2^n does not have a primitive root for $n \geq 3$. ∎

We have now narrowed down the possibilities for finding integers with primitive roots. They must be 2, 4, p^n, or $2p^n$ with p an odd prime. Naturally, we ask: Does every integer of this form have a primitive root? The answer is yes! The rest of this chapter is devoted to proving this result.

To begin with, 1 is a primitive root of 2, while 3 is a primitive root of 4. So far this is easy; to show that p^n has a primitive root is difficult. Our approach will be to first show that p has a primitive root and then, inductively, that p^n has a primitive root. Once we have shown this, it will be pure sport to show that $2p^n$ has a primitive root. In fact, to separate the easy proofs from the hard proofs (and so far all our proofs in this chapter fall into the "easy" category), let us make believe that we have already shown that p^n has a primitive root and use this fact to show that $2p^n$ has a primitive root.

Theorem 8.5. Let p be an odd prime and $n \geq 1$. If a is a primitive root of p^n, then either a or $a + p^n$ is a primitive root of $2p^n$.

Proof. We are assuming that p^n has a primitive root, call it a. We may, moreover, assume that a is odd, for if it is even, replace it with $a + p^n$, which is odd and is also a primitive root, since $a + p^n \equiv a \bmod p^n$. Now a is a primitive root of p^n, so $(a, p^n) = 1$. Also, a is odd, so $(2, a) = 1$. Thus $(a, 2p^n) = 1$ (this is why we needed to have a odd).

Let k be the order of a modulo $2p^n$. We claim that $k = \varphi(2p^n) = \varphi(2)\varphi(p^n) = p^{n-1}(p-1)$. Of course, $k \mid p^{n-1}(p-1)$, by Euler's Theorem. Now $a^k \equiv 1 \bmod(2p^n)$; hence $2p^n \mid a^k - 1$, so $p^n \mid a^k - 1$. Therefore, $a^k \equiv 1 \bmod p^n$. Because a is a primitive root of p^n, a has order $\varphi(p^n) = p^{n-1}(p-1)$, so $p^{n-1}(p-1) \mid k$ by Theorem 4.6. We conclude that $k = p^{n-1}(p-1) = \varphi(2p^n)$, completing the proof. ∎

So much for sport. We begin preparations to prove that powers of odd primes have primitive roots.

Lemma 8.6. Let $m > 1$ and let b and c be integers with $(b, m) = 1 = (c, m)$. If the order of b modulo m is k and the order of c modulo m is n and if $(k, n) = 1$, then the order of bc modulo m is kn.

Proof. Let s denote the order of bc modulo m. Since

$$(bc)^{kn} = (b^k)^n(c^n)^k \equiv 1^n \cdot 1^k = 1 \bmod m$$

we have that $s|kn$. We will now show that $k|s$ and $n|s$; since $(k, n) = 1$, this will show that $kn|s$ and hence that $kn = s$. By symmetry it is enough to show that $k|s$. We have $(bc)^s \equiv 1 \bmod m$; hence $(bc)^{sn} \equiv 1 \bmod m$. Since the order of c modulo m is n, $c^n \equiv 1 \bmod m$ hence $c^{sn} \equiv 1 \bmod m$. Then $1 \equiv (bc)^{sn} \equiv b^{sn} \cdot c^{sn} \equiv b^{sn} \cdot 1 \equiv b^{sn} \bmod m$. Therefore, by Theorem 4.6, $k|sn$. Since $(k, n) = 1$, we conclude that $k|s$. ∎

An easy induction yields:

Corollary 8.7. Let m be a positive integer and $b_1, \ldots, b_r$ integers such that $(b_i, m) = 1$ for $i = 1, 2, \ldots, r$. If b_i has order d_i modulo m and $(d_i, d_j) = 1$ for $i \neq j$, then $b_1 b_2 \cdots b_r$ has order $d_1 d_2 \cdots d_r$.

For a generalization of Lemma 8.6, see Exercise 8.B.5.

Theorem 8.8. Let p be a prime. Then p has a primitive root.

Proof. We seek an integer a such that $(a, p) = 1$ and the order of a modulo p is $\varphi(p) = p - 1$. Suppose that the prime decomposition of $p - 1$ is $p_1^{e_1} p_2^{e_2} \cdots p_r^{e_r}$. We will show that for each $i = 1, 2, \ldots, r$ there is an integer b_i whose order modulo p is $p_i^{e_i}$. From Corollary 8.7, $a = b_1 b_2 \cdots b_r$ will be the desired primitive root.

Let $1 \le i \le r$ and let $k_i = \dfrac{p-1}{p_i}$. We consider the congruence

$$X^{k_i} \equiv 1 \bmod p$$

By Theorem 6.19 this congruence has at most k_i solutions. Since $k_i < p - 1$, not all of the integers $1, 2, \ldots, p - 1$ are solutions of the congruence. Hence there is at least one integer $1 \le c_i \le p - 1$ that is not a solution to the congruence; that is, $c_i^{k_i} \not\equiv 1 \bmod p$. Let $l_i = p - 1/p_i^{e_i}$. Our choice for b_i is $c_i^{l_i}$. The aim is to show that $b_i \bmod p$ has order $p_i^{e_i}$.

For notational convenience, let $b = b_i$, $q = p_i$, $e = e_i$, $c = c_i$, and $l = l_i$. Finally, let s denote the order of b modulo p. Since

$$b^{q^e} = (c^l)^{q^e} = c^{p-1} \equiv 1 \bmod p$$

we have $s|q^e$. But q is prime, hence $s = q^f$ for some $0 \le f \le e$. Suppose $f < e$. Then

$$1 \equiv b^s = b^{q^f} = (c^l)^{q^f} \bmod p$$

Now $lq^f \cdot q^{e-f-1} = (p-1)/q = (p-1)/p_i = k_i$ hence

$$c^{k_i} \equiv 1 \bmod p,$$

contrary to $c = c_i$ not being a solution of $X^{k_i} \equiv 1 \bmod p$. Thus $f = e$ and the proof is completed. ∎

Theorem 8.9. If p is a prime, then p^2 has a primitive root.

Proof. Let a be a primitive root of p. We claim that one of a and $a + p$ is a primitive root of p^2. We let s be the order of a modulo p^2. Then $a^s \equiv 1 \bmod p^2$, so $p^2 | a^s - 1$; hence $p | a^s - 1$, showing that $a^s \equiv 1 \bmod p$. However, the order of a modulo p is $\varphi(p) = p - 1$, so that $p - 1 | s$. Since $p - 1$ and $p(p-1)$ differ only by a prime factor and $p - 1 | s$ while $s | p(p-1) = \varphi(p^2)$, clearly either $s = p - 1$ or $s = p(p-1)$. If $s = p(p-1)$, then a is a primitive root of p^2, since $\varphi(p^2) = p(p-1)$. Therefore, assume that $s = p - 1$ and consider $a + p$. Because $a + p \equiv a \bmod p$, $a + p$ is also a primitive root of p. The preceding argument applied to $a + p$ shows that the order of $a + p$ modulo p^2 is either $p - 1$ or $p(p-1)$. However, we are assuming that a has order $p - 1$ modulo p^2. Expanding $(a + p)^{p-1}$, we get

$$(a+p)^{p-1} = a^{p-1} + \binom{p-1}{1} a^{p-2} \cdot p + \cdots + \binom{p}{i} a^{p-1-i} p^i + \cdots + p^{p-1}$$

For $i \geq 2$, $p^i \equiv 0 \bmod p^2$. Thus

$$(a+p)^{p-1} \equiv a^{p-1} + \binom{p-1}{1} a^{p-2} \cdot p \equiv 1 + \binom{p-1}{1} a^{p-2} p \bmod p^2$$

Since $(a, p) = 1$ and $(p, p-1) = 1$, we have that $\binom{p-1}{1} a^{p-2} p \not\equiv 0 \bmod p^2$.

Thus $(a + p)^{p-1} \not\equiv 1 \bmod p^2$; hence $a + p$ has order $p(p-1)$ modulo p^2. ∎

The proof that powers of odd primes have primitive roots will now proceed by induction and will use arguments similar to those in Theorem 8.9. We will actually show that for an odd prime p and $n \geq 2$, if a is a primitive root of p^n, then a is also a primitive root of p^{n+1}.

Theorem 8.10. Let p be an odd prime and let $n \geq 1$. Then p^n has a primitive root.

Proof. We induct on n. The cases $n = 1$ and $n = 2$ are Theorems 8.8 and 8.9, respectively. Thus suppose that $n \geq 2$ and that we have shown that p^n has a primitive root, call it a. We will show that a is also a primitive root of p^{n+1}.

Let s be the order of a modulo p^{n+1}. Then s divides $\varphi(p^{n+1}) = p^n(p-1)$. Since $a^s \equiv 1 \bmod p^{n+1}$, we have $a^s \equiv 1 \bmod p^n$. The order of a modulo p^n is $\varphi(p^n) = p^{n-1}(p-1)$, so that we have $p^{n-1}(p-1)|s$. From $s|p^n(p-1)$ and $p^{n-1}(p-1)|s$ we conclude that either $s = p^{n-1}(p-1)$ or $s = p^n(p-1)$. We will eliminate the first possibility, thus showing that a has order $p^n(p-1) = \varphi(p^{n+1}) \bmod p^n$.

Surprisingly, we now proceed modulo p^{n-1}. Because $\varphi(p^{n-1}) = p^{n-2}(p-1)$, $a^{p^{n-2}(p-1)} \equiv 1 \bmod p^{n-1}$. We rewrite this congruence as $a^{p^{n-2}(p-1)} = 1 + kp^{n-1}$ for some integer k. We claim that $(p, k) = 1$. If not, then $p|k$, so $p^n|kp^{n-1}$. Thus $a^{p^{n-2}(p-1)} \equiv 1 \bmod p^n$. But this contradicts the fact that a is a primitive root of p^n and thus has order $p^{n-1}(p-1)$ modulo p^n. Therefore, $p \nmid k$, as claimed.

Returning to $a^{p^{n-2}(p-1)} = 1 + kp^{n-1}$, we raise both sides to the pth power to get

$$a^{p^{n-1}(p-1)} = 1 + pkp^{n-1} + \cdots + \binom{p}{i}(kp^{n-1})^i + \cdots + (kp^{n-1})^p$$

We examine each of these summands modulo p^{n+1}. For $i \geq 3$, we have $(n+1) \leq (n-1)i$ (this is not true for $i = 1$ or $i = 2$). Thus for $i \geq 3$, $p^{n+1} \Big| \binom{p}{i}(kp^{n-1})^i$. Hence

$$a^{p^{n-1}(p-1)} \equiv 1 + pkp^{n-1} + \binom{p}{2}(kp^{n-1})^2 \bmod p^{n+1}$$

However, $\binom{p}{2} = \dfrac{p(p-1)}{2}$, so that $\binom{p}{2}(kp^{n-1})^2 = \dfrac{p(p-1)}{2}k^2p^{2n-2} = \dfrac{p-1}{2}k^2p^{2n-1}$. Since $n + 1 \leq 2n - 1$, we see that p^{n+1} also divides $\binom{p}{2}(kp^{n-1})^2$. Therefore,

$$a^{p^{n-1}(p-1)} \equiv 1 + pkp^{n-1} \bmod p^{n+1}$$

Finally, $p \nmid k$, so $pkp^{n-1} \not\equiv 0 \bmod p^{n+1}$. Thus $a^{p^{n-1}(p-1)} \not\equiv 1 \bmod p^{n+1}$. As we saw in the second paragraph of the proof, this means that a has order $p^n(p-1) = \varphi(p^{n+1})$ modulo p^{n+1}, so a is a primitive root of p^{n+1}. This concludes the proof. ∎

Remark. We already know that $4 = 2^2$ does have a primitive root but that $8 = 2^3$ does not. Therefore, Theorem 8.10 fails for $p = 2$, $n = 2$. Where does the proof break down in this case? (Look closely at the statement p^{n+1} divides $\dfrac{p-1}{2}k^2p^{2n-1}$.)

There is one important question left unanswered. If we know that m has a primitive root, is there an easy way of finding it? No straightforward process is known; that is no process other than examining each of the powers of each integer relatively prime to m. As Theorems 8.5, 8.9, and 8.10 demonstrate, if a primitive root for an odd prime p is known, then it is easy to find primitive roots for p^n and $2p^n$. However, there is no easy way for finding primitive roots for primes. Theorem 8.8 only predicts the existence of a primitive root. This and related questions such as: "Many primes have 2 as a primitive root; are there infinitely many such primes?" await young and fertile minds.

EXERCISES 8.A

1. Find all primitive roots (if any) for each of the following integers: 5, 7, 10, 12.

2. Show that 3 is not a primitive root of 11, although it is a primitive root of 7.

3. Find a primitive root of 18 and verify Proposition 8.2.

4. Use the techniques illustrated after Proposition 8.2 to solve the following congruences.

(a) $5X \equiv 3 \bmod 14$ (b) $X^{11} \equiv 9 \bmod 14$
(c) $X^9 \equiv 5 \bmod 14$ (d) $X^9 \equiv 9 \bmod 14$

5. Let $m > 0$ and suppose that a is a primitive root of m. If $(b, m) = 1$, then there is a positive integer k such that $b \equiv a^k \bmod m$. The *least* such k is called *the index of b* (with respect to a and m) and is denoted by $\text{ind}_{a,m}(b)$. Calculate the following:

(a) $\text{ind}_{2,5}(3)$ (b) $\text{ind}_{2,5}(-121)$
(c) $\text{ind}_{3,14}(3)$ (d) $\text{ind}_{3,14}(13)$

6. Continuing with the notation in Exercise 8.A.5, show that the following statements are true for any b, c such that $(b, m) = (c, m) = 1$.

(a) $b \equiv c \mod m$ if and only if $\text{ind}_{a,m}(b) \equiv \text{ind}_{a,m}(c) \bmod \varphi(m)$.
(b) $\text{ind}_{a,m}(bc) \equiv \text{ind}_{a,m}(b) + \text{ind}_{a,m}(c) \bmod \varphi(m)$.
(c) $\text{ind}_{a,m}(b^t) \equiv t\, \text{ind}_{a,m}(b) \bmod \varphi(m)$.

Do you recognize any similarity between logarithms and indices?

7. In Exercise 8.A.1 you obtained 3 as a primitive root of 10. Observe that 7 is the multiplicative inverse of 3 modulo 10 and 7 is also a primitive root of 10. State and prove a general result.

8. Verify Theorem 8.5 by finding an odd primitive root of 7 and checking that it is also an odd primitive root of 14. Do the same for 11 and 22 as well as 13 and 26.

9. Let a be a primitive root of m. Show a^n is a primitive root of m if and only if $(n, \varphi(m)) = 1$.
10. Show that if m has a primitive root, then it has exactly $\varphi(\varphi(m))$ primitive roots between 0 and m.
11. Show that 18 is not a primitive root of $25 = 5^2$, although it is a primitive root of 5. Use 18 to find a primitive root of 25. (*Hint*: Examine the proof of Theorem 8.9.)
12. Assume that a is a primitive root of p^2, where p is a prime. Show by examples that $a + p$ may or may not be a primitive root of p^2.
13. Find a primitive root of 25 and verify that it is a primitive root of 125, a fact predicted by the first paragraph of the proof of Theorem 8.10.
14. Prove: If a is a primitive root of the prime p, $p \neq 2$, then either a is a primitive root of p^n for all $n \geq 1$ or $a + p$ is a primitive root of p^n for all $n \geq 1$.

EXERCISES 8.B

1. Suppose that m has a primitive root and let n be a positive integer with $(n, \varphi(m)) = 1$. Show that if c is any integer relatively prime to m, then the congruence $X^n \equiv c \bmod m$ has a solution.
2. In Exercise 8.B.1, how many solutions are there modulo m?
3. Assume that m has a primitive root and let c be relatively prime to m. If $n \geq 1$ and $d = (n, \varphi(m))$, show that the congruence $X^n \equiv c \bmod m$ either has no solutions or else has exactly d solutions modulo m. In the second case describe the pattern the solutions form, in terms of the powers of the primitive root.
4. Discuss how the proof of Theorem 8.5 was made easy by the fact that $\varphi(2m) = \varphi(m)$ whenever m is an odd integer greater than 1.
5. Let $m > 0$ and suppose that $(b, m) = (c, m) = 1$. Show that if the order of b modulo m is k and the order of c modulo m is h, then the order of bc modulo m is $[h, k]$ = the least common multiple of h and k.
6. Let p be a prime and let $p - 1 = q^t m$, where q is a prime and $(q, m) = 1$. Let $c \in \mathbb{Z}$ with $(p, c) = 1$ and let $b = c^m$. Show that the order of b modulo p divides q^t. What restriction on c ensures that the order of b is q^t? (*Hint*: See the proof of Theorem 8.8.)
7. Using the notation of Exercise 8.B.6, show that if the order of b modulo p is q^t, then c is not a solution of the congruence $X^{(p-1)/q} \equiv 1 \bmod p$.
8. Prove Wilson's Theorem (Theorem 4.19) using the fact that every prime has a primitive root.
9. Let a be a primitive root of the prime p. Show that the set $\{a^2, a^4, \ldots, a^{p-1}\}$ represents a complete set of quadratic residues modulo p.

9

Quadratic Reciprocity

WE now return to the problem of solving the congruence

$$(*) \qquad X^2 \equiv a \bmod p, \qquad p \text{ a prime}$$

You may recall from Chapter 6 that this type of congruence was arrived at through a reduction process which originally began with an attempt to solve the congruence

$$X^2 \equiv a \bmod m, \qquad m > 0$$

An obvious way to find solutions to $(*)$ is simply to search through the numbers $1^2, 2^2, 3^2, \ldots, (p-1)^2$ and determine if any of these are congruent modulo p to a. The drawbacks are painfully clear—large primes can lead to large headaches. What we will present here is a method for determining if solutions exist, the key result being the stunning Quadratic Reciprocity Theorem (Theorem 9.6). Full appreciation can only be achieved after you have used this magnificent result.

9.1 The Legendre Symbol

We will assume that $(a, p) = 1$; otherwise, we have the congruence $X^2 \equiv 0 \bmod p$ and its solution is $\{x \mid x \equiv 0 \bmod p\}$. Second, we will suppose that $p \neq 2$, because if $p = 2$, then there is always a solution. Thus for $p > 2$, we are in effect asking which of the integers $1, 2, 3, \ldots, p-1$ are quadratic residues mod p.

Lemma 9.1. Let $p > 2$ be a prime. Then there are exactly $\dfrac{p-1}{2}$ distinct nonzero quadratic residues mod p.

Proof. First note that p is odd so that $\dfrac{p-1}{2}$ is an integer. Obviously, all of the nonzero quadratic residues mod p occur in the list

$$1^2, 2^2, 3^2, \ldots, (p-1)^2$$

so we must only account for repetitions mod p in order to count the distinct ones. Since $1^2 \equiv (p-1)^2 \bmod p$, $2^2 \equiv (p-2)^2 \bmod p$, and so on, we see that the second half of our list taken mod p simply repeats the first half (in reverse order). Thus we need only consider the first half,

$$1^2, 2^2, \ldots, \left(\frac{p-1}{2}\right)^2$$

We now claim that these are all distinct mod p. For if $1 \leq i \leq \frac{p-1}{2}$ and $1 \leq j \leq \frac{p-1}{2}$ and $i^2 \equiv j^2$, then both i and j solve $X^2 \equiv i^2 \bmod p$. By Exercise 4.B.16, $i = j$, as desired. Thus $1^2, 2^2, \ldots, \left(\frac{p-1}{2}\right)^2$ give all the distinct nonzero quadratic residues mod p, and we are done. ∎

Example. With $p = 13$, we have $\frac{p-1}{2} = 6$ and

$$1^2 \equiv 1 \equiv (12)^2; \quad 2^2 \equiv 4 \equiv (11)^2; \quad 3^2 \equiv 9 \equiv (10)^2;$$
$$4^2 \equiv 3 \equiv (9)^2; \quad 5^2 \equiv 12 \equiv 8^2; \quad 6^2 \equiv 10 \equiv 7^2 \pmod{13}$$

so the six quadratic residues of 13 which are relatively prime to 13 are

$$1, 4, 9, 3, 12, 10$$

Our next result tells us how quadratic residues behave under multiplication.

Theorem 9.2. Let $p > 2$ be a prime and $a, b \in \mathbb{Z}$ with $(a, p) = (b, p) = 1$.

(i) If a and b are quadratic residues mod p, then ab is a quadratic residue mod p.

(ii) If a is a quadratic residue mod p and b is a quadratic nonresidue mod p, then ab is a quadratic nonresidue mod p.

(iii) If a and b are quadratic nonresidues mod p, then ab is a quadratic residue mod p.

Proof.

(i) Suppose that $x, y \in \mathbb{Z}$ such that $x^2 \equiv a \bmod p$ and $y^2 \equiv b \bmod p$. Then $(xy)^2 \equiv x^2y^2 \equiv ab \bmod p$.

(ii) Let $x \in \mathbb{Z}$ such that $x^2 \equiv a \bmod p$ and assume that ab is a quadratic residue mod p: say $z^2 \equiv ab \bmod p$. We will show that b must then be a quadratic residue mod p. Since $(a, p) = 1$, we have $(x^2, p) = 1$ by

Lemma 2.15, so $(x, p) = 1$. Thus by Theorem 4.3, there exists $y \in \mathbb{Z}$ such that $xy \equiv 1 \bmod p$. Then

$$z^2 \equiv ab \equiv x^2 b \bmod p$$

so

$$y^2 z^2 \equiv y^2 x^2 b \equiv (yx)^2 b \equiv b \bmod p$$

That is,

$$(yz)^2 \equiv b \bmod p$$

Hence if a is a quadratic residue mod p and b is a quadratic nonresidue mod p, then necessarily ab cannot be a quadratic residue mod p.

(iii) We know that there are exactly $\dfrac{p-1}{2}$ distinct nonzero quadratic residues mod p; denote these by $c_1, c_2, \ldots, c_{\left(\frac{p-1}{2}\right)}$. Since we are assuming that a is not a quadratic residue mod p, by (ii) we know that $ac_1, ac_2, \ldots, ac_{\left(\frac{p-1}{2}\right)}$ are all quadratic nonresidues mod p. Since $(a, p) = 1$, by Theorem 4.2 these $\dfrac{p-1}{2}$ integers must be distinct modulo p. But there are exactly $\dfrac{p-1}{2}$ distinct quadratic *nonresidues* mod p and hence $ac_1, ac_2, \ldots, ac_{\left(\frac{p-1}{2}\right)}$ must be all of them. If ab is a quadratic nonresidue mod p, then $ab \equiv ac_i \bmod p$, for some $1 \le i \le \left(\dfrac{p-1}{2}\right)$. But then $b \equiv c_i$ because $(a, p) = 1$, so b is a quadratic residue mod p; this contradicts the assumption that b is a quadratic nonresidue mod p. Thus ab is a quadratic residue mod p. ∎

The next definition gives what appears to be an unusual use of notation. Usually, notation is introduced to express a complicated idea in a simple fashion. The symbol we are about to define will always represent either $+1$ or -1. Thus at first sight it appears that our notation is a difficult way of expressing ± 1. However, you will see that the symbol acts as a bridge between the sophisticated idea of quadratic residues and the simple idea of ± 1. The symbol is called the *Legendre symbol.* The symbol was introduced by none other than A. M. Legendre (1752–1833), who rediscovered the Quadratic Reciprocity Theorem in 1785 after Euler, who had found it in 1783. As in Euler's case, Legendre failed to prove it, but he left behind (among other things) a useful idea.

Definition. Let $p > 2$ be a prime and $a \in \mathbb{Z}$ with $(a, p) = 1$. We let $\left(\frac{a}{p}\right) = 1$ if a is a quadratic residue mod p, and $\left(\frac{a}{p}\right) = -1$ if a is a quadratic nonresidue mod p.

As examples we have $\left(\frac{1}{5}\right) = 1$; $\left(\frac{3}{11}\right) = 1$, since $6^2 \equiv 3 \bmod 11$; $\left(\frac{6}{7}\right) = -1$, since the quadratic residues mod 7 are 1, 2, and 4.

Here are some basic properties about the Legendre symbol $\left(\frac{a}{p}\right)$.

Theorem 9.3. Let $p > 2$ be a prime and $a, b \in \mathbb{Z}$ with $(a, p) = (b, p) = 1$.

(i) If $a \equiv b \bmod p$, then $\left(\frac{a}{p}\right) = \left(\frac{b}{p}\right)$.

(ii) $\left(\frac{a^2}{p}\right) = 1$.

(iii) $\left(\frac{ab}{p}\right) = \left(\frac{a}{p}\right)\left(\frac{b}{p}\right)$.

Proof.

(i) There are two cases to consider. Suppose that $\left(\frac{a}{p}\right) = 1$. Then there is an $x \in \mathbb{Z}$ such that $x^2 \equiv a \bmod p$. Since $a \equiv b \bmod p$, we then have $x^2 \equiv b \bmod p$. Therefore, b is a quadratic residue mod p, so $\left(\frac{b}{p}\right) = 1 = \left(\frac{a}{p}\right)$. Next, suppose that $\left(\frac{a}{p}\right) = -1$; then there are no solutions to the congruence $X^2 \equiv a \bmod p$. As $a \equiv b \bmod p$, there can be no solutions to $X^2 \equiv b \bmod p$, either. Thus $\left(\frac{b}{p}\right) = -1 = \left(\frac{a}{p}\right)$.

(ii) We have a as a solution of $X^2 \equiv a^2 \bmod p$, so a^2 is a quadratic residue mod p. Thus $\left(\frac{a^2}{p}\right) = 1$ by definition.

(iii) This is really a consequence of Theorem 9.2. If $\left(\frac{a}{p}\right) = \left(\frac{b}{p}\right) = 1$, then a and b are both quadratic residues mod p, so by Theorem 9.2, ab is a quadratic residue mod p. Hence $\left(\frac{ab}{p}\right) = 1 = 1 \cdot 1 = \left(\frac{a}{p}\right) \cdot \left(\frac{b}{p}\right)$. The remaining cases will be left as an exercise (Exercise 9.A.4). ∎

As we mentioned in the proof, (iii) of Theorem 9.3 is nothing more than Theorem 9.2 in disguise. It is precisely because it is such an efficient disguise that the Legendre symbol is so effective. [The reader familiar with the terms will recognize that the Legendre symbol is a "homomorphism" between the multiplicative group $\{a \in \mathbb{Z} \mid 1 \le a \le p-1 \text{ and } (a, p) = 1\}$ with multiplication modulo p and the multiplicative group $\{1, -1\}$.]

Since $1^2 = 1$, we have that $\left(\frac{1}{p}\right) = 1$ for all primes $p > 2$. On the other hand, $\left(\frac{-1}{5}\right) = 1$ since $2^2 \equiv -1 \bmod 5$, but $\left(\frac{-1}{7}\right) = -1$ because the only quadratic residues mod 7 are 1, 4, and 2, none of which are congruent to -1 modulo 7. We may ask, then, for what primes p is -1 a quadratic residue mod p? Let us tabulate a few values of $\left(\frac{-1}{p}\right)$.

p	$\left(\frac{-1}{p}\right)$	Reason
3	-1	Quadratic residues are: 1.
5	1	$2^2 \equiv -1 \bmod 5$
7	-1	Quadratic residues are: 1, 4, 2.
11	-1	Quadratic residues are: 1, 4, 9, 5, 3.
13	1	$5^2 \equiv -1 \bmod 13$
17	1	$4^2 \equiv -1 \bmod 17$

Now observe that if p is any prime, $p > 2$, then $p \equiv 1 \bmod 4$ or $p \equiv 3 \bmod 4$, since p is odd. Also, 5, 13, and 17 are congruent to 1 modulo 4, while 3, 7, and 11 are congruent to 3 modulo 4. With this evidence the next result should come as no surprise. In fact, it is just an old friend in new clothing.

Theorem 9.4. Let $p > 2$ be a prime. If $p \equiv 1 \bmod 4$, then $\left(\frac{-1}{p}\right) = 1$, while if $p \equiv 3 \bmod 4$, then $\left(\frac{-1}{p}\right) = -1$.

Proof. $\left(\frac{-1}{p}\right) = 1$ means that there is an $x \in \mathbb{Z}$ with $x^2 \equiv -1 \bmod p$. Thus $\left(\frac{-1}{p}\right) = 1$ exactly when $X^2 \equiv -1 \bmod p$ has a solution. The present result is therefore just another form of Theorem 4.20. ∎

Corollary 9.5. Let $p \neq 2$ be a prime and suppose that $(a, p) = 1$. If $p \equiv 1 \bmod 4$, then $\left(\frac{a}{p}\right) = \left(\frac{p-a}{p}\right)$. If $p \equiv 3 \bmod 4$, then $\left(\frac{a}{p}\right) = -\left(\frac{p-a}{p}\right)$.

Proof. Since $p - a \equiv -a \bmod p$, Theorem 9.2 tells us that $\left(\frac{p-a}{p}\right) = \left(\frac{-a}{p}\right) = \left(\frac{-1}{p}\right)\left(\frac{a}{p}\right)$. Thus, if $p \equiv 1 \bmod 4$, then $\left(\frac{-1}{p}\right) = 1$, so $\left(\frac{p-a}{p}\right) = \left(\frac{a}{p}\right)$, while if $p \equiv 3 \bmod 4$, then $\left(\frac{-1}{p}\right) = -1$, so $\left(\frac{p-a}{p}\right) = -\left(\frac{a}{p}\right)$. ∎

What we have learned so far about the Legendre symbol is at best of minor help in trying to determine quadratic residues. Thus if we wish to know if $\left(\frac{1678}{2503}\right)$ has value 1 or -1 (note: 2503 is a prime), we might start by using Corollary 9.5. Since $2503 - 1678 = 825$ and $2503 \equiv 3 \bmod 4$, we have

$$\left(\frac{1678}{2503}\right) = -\left(\frac{825}{2503}\right)$$

Furthermore, $825 = 3 \cdot 5^2 \cdot 11$; hence by Theorem 9.3,

$$\left(\frac{1678}{2503}\right) = -\left(\frac{825}{2503}\right) = -\left(\frac{3}{2503}\right)\left(\frac{5^2}{2503}\right)\left(\frac{11}{2503}\right)$$

By (ii) of Theorem 9.3, $\left(\frac{5^2}{2503}\right) = 1$, so

$$\left(\frac{1678}{2503}\right) = -\left(\frac{3}{2503}\right)\left(\frac{11}{2503}\right)$$

This, however, is about as far as our present knowledge will take us. While we might feel more comfortable with the smaller numbers 3 and 11, frankly it is not much easier to calculate $\left(\frac{3}{2503}\right)$, then $\left(\frac{1678}{2503}\right)$.

9.2 The Quadratic Reciprocity Theorem

What we have so far, then, are techniques for manipulating the top number in the Legendre symbol. What is needed is a technique for operating with the bottom number. That is exactly what quadratic reciprocity provides. We now state this lovely theorem and demonstrate how to use it. We will then devote the remainder of this chapter to proving it.

Theorem 9.6 (Quadratic Reciprocity). Let p and q be distinct odd primes. If $p \equiv 3 \bmod 4$ and $q \equiv 3 \bmod 4$, then $\left(\frac{p}{q}\right) = -\left(\frac{q}{p}\right)$. Otherwise, $\left(\frac{p}{q}\right) = \left(\frac{q}{p}\right)$.

As an example, let $p = 13$ and $q = 17$. Since $13 \not\equiv 3 \bmod 4$, Quadratic Reciprocity claims that $\left(\frac{13}{17}\right) = \left(\frac{17}{13}\right)$. We can check this out directly:

$$8^2 \equiv 13 \bmod 17 \qquad \text{and} \qquad 17 \equiv 4 \equiv 2^2 \bmod 13$$

so $\left(\frac{13}{17}\right) = 1 = \left(\frac{17}{13}\right)$.

As another example, take $p = 7$ and $q = 11$. Here $7 \equiv 3 \bmod 4$ and $11 \equiv 3 \bmod 4$; Quadratic Reciprocity claims that $\left(\frac{7}{11}\right) = -\left(\frac{11}{7}\right)$. To verify this directly, we first find, by checking the squares $1^2, 2^2, \ldots, (10)^2$, that 7 is not a quadratic residue mod 11, so $\left(\frac{7}{11}\right) = -1$. On the other hand, $11 \equiv 4 \equiv 2^2 \bmod 7$, so $\left(\frac{11}{7}\right) = 1$. Thus $-\left(\frac{11}{7}\right) = -1 = \left(\frac{7}{11}\right)$.

Before proving Theorem 9.6, we wish to show you how we can calculate $\left(\frac{a}{p}\right)$ for any prime $p > 2$ and any $a \in \mathbb{Z}$ for which $(a, p) = 1$. We start with some easy examples, before going on to more complicated examples.

EXAMPLES.

(a) We consider $\left(\frac{7}{11}\right)$. Both 7 and 11 are odd primes, $7 \equiv 3 \bmod 4$ and $11 \equiv 3 \bmod 4$, so by 9.6, $\left(\frac{7}{11}\right) = -\left(\frac{11}{7}\right)$. However, $11 \equiv 4 \bmod 7$, so $\left(\frac{11}{7}\right) = \left(\frac{4}{7}\right)$ by Theorem 9.3. However, $4 = 2^2$, so again by Theorem 9.3, $\left(\frac{4}{7}\right) = 1$. Thus

$$\left(\frac{7}{11}\right) = -\left(\frac{11}{7}\right) = -\left(\frac{4}{7}\right) = -1$$

(b) Next consider $\left(\frac{15}{31}\right)$. First, 15 is not prime, so Quadratic Reciprocity cannot be used at once. However, $15 = 3 \cdot 5$, so by Theorem 9.3, $\left(\frac{15}{31}\right) = \left(\frac{3}{31}\right)\left(\frac{5}{31}\right)$. Thus we have to evaluate the two symbols $\left(\frac{3}{31}\right)$ and $\left(\frac{5}{31}\right)$. Let us take $\left(\frac{3}{31}\right)$ first. Here $3 \equiv 3 \bmod 4$ and $31 \equiv 3 \bmod 4$, so by Quadratic Reciprocity, $\left(\frac{3}{31}\right) = -\left(\frac{31}{3}\right)$. Since $31 \equiv 1 \bmod 3$ and $\left(\frac{1}{3}\right) = 1$, we then have $\left(\frac{3}{31}\right) = -\left(\frac{1}{3}\right) = -1$. Next we look at $\left(\frac{5}{31}\right)$; $5 \equiv 1 \bmod 4$, so by Theorem 9.6, $\left(\frac{5}{31}\right) = \left(\frac{31}{5}\right)$. Also $31 \equiv 1 \bmod 5$, so $\left(\frac{5}{31}\right) = \left(\frac{1}{5}\right) = 1$. Combining our two results we obtain $\left(\frac{15}{31}\right) = \left(\frac{3}{31}\right)\left(\frac{5}{31}\right) = -1 \cdot 1 = -1$. Thus 15 is not a quadratic nonresidue mod 31, so the congruence $X^2 \equiv 15 \bmod 31$ has no solutions.

(c) We will calculate $\left(\frac{11}{13}\right)$, leaving you to fill in the appropriate reasons. First, $13 \equiv 1 \bmod 4$, so $\left(\frac{11}{13}\right) = \left(\frac{13}{11}\right) = \left(\frac{2}{11}\right)$. Here 2 is not an odd prime, so Quadratic Reciprocity cannot be applied. However, $11 \equiv 3 \bmod 4$, so $\left(\frac{2}{11}\right) = -\left(\frac{11-2}{11}\right)$ (remember 9.5); hence $\left(\frac{2}{11}\right) = -\left(\frac{9}{11}\right) = -\left(\frac{3^2}{11}\right) = -1 \cdot 1 = -1$. Accordingly, $\left(\frac{11}{13}\right) = -1$.

(d) For our final example we will calculate $\left(\frac{595}{8243}\right)$. To facilitate things for you, we collect and summarize the properties of the Legendre symbol which we will use. In (i) through (vii) below, p and q designate distinct odd primes, a, b are integers, both relatively prime to p.

(i) If $a \equiv b \bmod p$, then $\left(\frac{a}{p}\right) = \left(\frac{b}{p}\right)$.

(ii) $\left(\frac{a^2}{p}\right) = 1$.

(iii) $\left(\frac{ab}{p}\right) = \left(\frac{a}{p}\right)\left(\frac{b}{p}\right)$.

(iv) If $p \equiv 1 \bmod 4$, then $\left(\frac{a}{p}\right) = \left(\frac{p-a}{p}\right)$.

(v) If $p \equiv 3 \bmod 4$, then $\left(\frac{a}{p}\right) = -\left(\frac{p-a}{p}\right)$.

(vi) If $p \equiv 3 \bmod 4$ and $q \equiv 3 \bmod 4$, then $\left(\frac{p}{q}\right) = -\left(\frac{q}{p}\right)$.

(vii) If either $p \equiv 1 \bmod 4$ *or* $q \equiv 1 \bmod 4$, then $\left(\frac{p}{q}\right) = \left(\frac{q}{p}\right)$.

We now proceed to evaluate $\left(\frac{595}{8243}\right)$; you should, of course, note in this as well as other examples, that the bottom number is an odd prime; otherwise, the Legendre symbol as we have defined it is meaningless. Now $595 = 5 \cdot 7 \cdot 17$, so $\left(\frac{595}{8243}\right) = \left(\frac{5}{8243}\right)\left(\frac{7}{8243}\right)\left(\frac{17}{8243}\right)$, by (iii).

We first attack $\left(\frac{5}{8243}\right)$. Since $5 \equiv 1 \bmod 4$, we have by (vii) that $\left(\frac{5}{8243}\right) = \left(\frac{8243}{5}\right)$. Then $8243 \equiv 3 \bmod 5$, so $\left(\frac{5}{8243}\right) = \left(\frac{8243}{5}\right) = \left(\frac{3}{5}\right)$. Now $5 \equiv 1 \bmod 4$, so by (vii), $\left(\frac{3}{5}\right) = \left(\frac{5}{3}\right) = \left(\frac{2}{3}\right)$ because $5 \equiv 2 \bmod 3$. Finally, $3 \equiv 3 \bmod 4$, so (v) gives $\left(\frac{2}{3}\right) = -\left(\frac{3-2}{3}\right) = -\left(\frac{1}{3}\right) = -1$. Thus $\left(\frac{5}{8243}\right) = -1$.

Next consider $\left(\frac{7}{8243}\right)$; $7 \equiv 3 \bmod 4$ and $8243 \equiv 3 \bmod 4$, so that $\left(\frac{7}{8243}\right) = -\left(\frac{8243}{7}\right)$. Since $8243 \equiv 4 \bmod 7$, we have $\left(\frac{7}{8243}\right) = -\left(\frac{4}{7}\right)$ $= -\left(\frac{2^2}{7}\right) = -1 \cdot 1 = -1$.

We have left $\left(\frac{17}{8243}\right)$. Here $17 \equiv 1 \bmod 4$, so by (vii), $\left(\frac{17}{8243}\right) = \left(\frac{8243}{17}\right) = \left(\frac{15}{17}\right)$ because $8243 \equiv 15 \bmod 17$. At this point we might be tempted to use (iv) because $17 - 15 = 2$; however, 2 is not covered by (vi) or (vii), so this step would not be advisable (unless we wish to check out the squares from 1 to 16). Instead, we have $\left(\frac{15}{17}\right) = \left(\frac{3}{17}\right)\left(\frac{5}{17}\right)$. Then $17 \equiv 1 \bmod 4$, so $\left(\frac{3}{17}\right) = \left(\frac{17}{3}\right) = \left(\frac{2}{3}\right) = -\left(\frac{3-2}{3}\right) = -\left(\frac{1}{3}\right) = -1$. Also, $\left(\frac{5}{17}\right) = \left(\frac{17}{5}\right) = \left(\frac{2}{5}\right) = \left(\frac{5-2}{5}\right)$ [by (iv)] $= \left(\frac{3}{5}\right) = \left(\frac{5}{3}\right)$ [by (vii)] $= \left(\frac{2}{3}\right) = -\left(\frac{1}{3}\right)$ [by (v)] $= -1$. Thus $\left(\frac{17}{8243}\right) = \left(\frac{15}{17}\right) = \left(\frac{3}{17}\right)\left(\frac{5}{17}\right) = (-1)(-1) = 1$.

Finally, we combine these three results to obtain

$$\left(\frac{595}{8243}\right) = \left(\frac{5}{8243}\right)\left(\frac{7}{8243}\right)\left(\frac{17}{8243}\right) = (-1)(-1)(1) = 1$$

We wish to make it clear that the examples given above are only intended to show a method by which $\left(\frac{a}{p}\right)$ can be evaluated. We are not claiming that in many cases there might not be a quicker way. For instance in one step of example (d), we were confronted with $\left(\frac{3}{17}\right)$ and proceeded to evaluate it using the method. It would have probably been just as quick to check out the squares $1^2, 2^2, 3^2, \ldots, 8^2$ $\left(8 = \frac{17-1}{2}\right)$ and come out with the same result of -1. We deliberately avoided shortcuts, so as to better illustrate the general method. There is one shortcut, however, which deserves mention. When faced with $\left(\frac{2}{p}\right)$, our method is to replace it by $\pm\left(\frac{p-2}{p}\right)$, using (*v*). Actually, there is a result that gives $\left(\frac{2}{p}\right)$ almost instantly. We will simply state the theorem and outline its proof as exercises (see Exercise 9.B.2–5).

Theorem 9.7. Let p be an odd prime. Then

$$\left(\frac{2}{p}\right) = 1 \qquad \text{if} \quad p \equiv \pm 1 \bmod 8$$

and

$$\left(\frac{2}{p}\right) = -1 \qquad \text{if} \quad p \equiv \pm 3 \bmod 8$$

It is now time to begin the proof of the Quadratic Reciprocity Theorem. A first step is to restate Theorem 6.19 for your convenience.

Theorem 9.8. The congruence

$$a_n X^n + a_{n-1} X^{n-1} + \cdots + a_1 X + a_0 \equiv 0 \bmod p$$

has at most n distinct solutions modulo p.

Next recall that in Exercise 4.B.6 you were asked to show that for a prime $p > 2$ if $(a, p) = 1$, then either

$$a^{\frac{p-1}{2}} \equiv 1 \bmod p$$

or

$$a^{\frac{p-1}{2}} \equiv -1 \bmod p$$

Here is one way of solving that exercise which uses Theorem 9.8. The congruence

$$X^2 \equiv 1 \bmod p$$

has at most two distinct solutions modulo p. Since $1^2 \equiv 1 \bmod p$ and $(-1)^2 \equiv 1 \bmod p$, the only solutions modulo p are 1 and -1. By Fermat's Theorem, if $(a, p) = 1$ and $p < 2$,

$$\left(a^{\frac{p-1}{2}}\right)^2 \equiv a^{p-1} \equiv 1 \bmod p$$

so $a^{\frac{p-1}{2}} \equiv 1 \bmod p$ or $a^{\frac{p-1}{2}} \equiv -1 \bmod p$.

We now relate this observation to quadratic residues modulo p.

Theorem 9.9 (Euler's Criterion). Let $p > 2$ be a prime and let $a \in \mathbb{Z}$ with $(a, p) = 1$. Then

$$\left(\frac{a}{p}\right) \equiv a^{\frac{p-1}{2}} \bmod p$$

Proof. Suppose first that $\left(\frac{a}{p}\right) = 1$. Then a is a quadratic residue mod p, so there is an integer b such that

$$b^2 \equiv a \bmod p$$

Then

$$a^{\frac{p-1}{2}} \equiv (b^2)^{\frac{p-1}{2}} \equiv b^{p-1} \equiv 1 \bmod p$$

Thus

$$\left(\frac{a}{p}\right) = 1 \equiv a^{\frac{p-1}{2}} \bmod p$$

This also shows that every nonzero quadratic residue mod p is a solution of the congruence

$$X^{\frac{p-1}{2}} \equiv 1 \bmod p$$

By Lemma 9.1 there are exactly $\frac{p-1}{2}$ distinct nonzero quadratic residues mod p. But we know from Theorem 9.8 that the congruence

$$X^{\frac{p-1}{2}} \equiv 1 \bmod p$$

has at most $\frac{p-1}{2}$ distinct solutions modulo p. Therefore, the set of all distinct nonzero quadratic residues accounts for all the solutions to this congruence. Now if $\left(\frac{a}{p}\right) = -1$, then a is a quadratic nonresidue mod p, so a cannot be a solution to the congruence $X^{\frac{p-1}{2}} \equiv 1 \bmod p$. Thus $a^{\frac{p-1}{2}} \not\equiv 1 \bmod p$, which means then that $a^{\frac{p-1}{2}} \equiv -1 \bmod p$ as a result of the discussion preceding this theorem. Hence $\left(\frac{a}{p}\right) = -1 \equiv a^{\frac{p-1}{2}} \bmod p$. ∎

Euler's Criterion is of little use for computing Legendre symbols directly. For instance, in order to use it to evaluate $\left(\frac{595}{8243}\right)$, we must determine if $(595)^{4121}$ is congruent to 1 or to -1 modulo 8243. This is a formidable calculation. The value of Euler's Criterion lies in its theoretical nature and it will play a crucial role in the proof of the Quadratic Reciprocity Theorem. To show you the strength of Euler's Criterion, we mention that both Theorem 9.3(iii) and Theorem 9.4 are consequences of it. In Exercise 9.B.1 you are asked to verify this claim.

Our proof of the Quadratic Reciprocity Theorem will follow an outline suggested by one of Gauss' many proofs of this theorem. We have incorporated some simplifications which are adapted from a recent article by J. S. Frame. The article, "A Short Proof of Quadratic Reciprocity," appeared in the *American Mathematical Monthly*, Vol. 85 (1978), pp. 818–819.

The first step will require establishing notation which will be used in the remainder of the chapter. Hence let p be a prime, $p > 2$, and let $a \in \mathbb{Z}$ such that $(a, p) = 1$. [For the moment we only need $(a, p) = 1$; later we will impose the additional requirement that a be odd also.] We consider the multiples of a: $a, 2 \cdot a, 3 \cdot a, \ldots, \left(\frac{p-1}{2}\right) \cdot a$; there are exactly $\frac{p-1}{2}$ of these multiples. Now for $k = 1, 2, \ldots, \frac{p-1}{2}$, we express $k \cdot a$ as

$$k \cdot a = p \cdot m_k + r_k, \qquad \text{where} \quad 1 \le r_k < p$$

(*Question*: Why is $r_k = 0$ not possible?) We will now take the set of remainders $\left\{r_k \mid 1 \le k \le \frac{p-1}{2}\right\}$ and split this set into two sets S and T. The set S will consist of those remainders which are less than or equal to $\frac{p-1}{2}$, and the set T will consist of those remainders which are greater than $\frac{p-1}{2}$. Symbolically:

$$S = \left\{r_k \mid 1 \le k \le \frac{p-1}{2} \quad \text{and} \quad 1 \le r_k \le \frac{p-1}{2}\right\}$$

$$T = \left\{r_k \mid 1 \le k \le \frac{p-1}{2} \quad \text{and} \quad \frac{p-1}{2} < r_k < p\right\}$$

Finally, we let U be the set

$$U = \{p - r_k \mid r_k \in T\}$$

and we let $t =$ the number of elements in the set T. This is the notation that will be used throughout, and it is worthwhile to look at a particular example at this point.

EXAMPLE. We select $p = 17$ and $a = 5$ so that $\frac{p-1}{2} = 8$. In constructing the remainders r_k, we will have to let k assume the values $k = 1, 2, \ldots, 8$. You can verify that the following table is correct:

k	1	2	3	4	5	6	7	8
$k \cdot a = k \cdot 5$	5	10	15	20	25	30	35	40
m_k	0	0	0	1	1	1	2	2
r_k	5	10	15	3	8	13	1	6

For example, with $k = 4$, $k \cdot a = 4 \cdot 5 = 20$ and $20 = 1 \cdot 17 + 3$, so that $m_4 = 1$ and $r_4 = 3$. What are the sets S, T, and U? We have

$$S = \{r_k \mid 1 \le k \le 8 \quad \text{and} \quad 1 \le r_k \le 8\} = \{r_1, r_4, r_5, r_7, r_8\} = \{5, 3, 8, 1, 6\}$$

$$T = \{r_k \mid 1 \le k \le 8 \quad \text{and} \quad 8 < r_k < 17\} = \{r_2, r_3, r_6\} = \{10, 15, 13\}$$

$$U = \{17 - r_k \mid r_k \in T\} = \{17 - 10, 17 - 15, 17 - 13\} = \{7, 2, 4\}$$

Last of all, t(= number of elements in T), is, of course, 3.

You might linger a bit and examine how these sets interact. In the example, $S \cup T = \{5, 10, 15, 3, 8, 13, 1, 6\}$ while $S \cap T = \varnothing$. You should easily see that this is true in general; that is, $S \cup T = \left\{r_k \mid 1 \leq k \leq \frac{p-1}{2}\right\}$ and $S \cap T = \varnothing$ for any prime $p > 2$ and any integer a relatively prime to p. Another observation concerning the example is that $S \cup U = \{1, 2, \ldots, 8\}$ while $S \cap U = \varnothing$. In general, these relationships hold, although they are not as immediately obvious as the relationships between S and T. Let's prove it.

Lemma 9.10. $S \cap U = \varnothing$ and $S \cup U = \left\{1, 2, \ldots, \frac{p-1}{2}\right\}$.

Proof. To show that $S \cap U = \varnothing$, let us assume that $x \in S \cap U$ and derive a contradiction. Since $x \in S$, $x = r_i$ with $1 \leq i \leq \frac{p-1}{2}$. Also, $x \in U$, so $x = p - r_j$ for some $1 \leq j \leq \frac{p-1}{2}$. Now for $k = 1, 2, \ldots, \frac{p-1}{2}$, we have $k \cdot a = m_k p + r_k$; hence $k \cdot a \equiv r_k \bmod p$. Applying this to $r_i = p - r_j$ and r_j, we have

$$ia \equiv r_i \equiv p - r_j \equiv -r_j \equiv -ja \bmod p$$

so $(i + j)a \equiv 0 \bmod p$. Since $(a, p) = 1$, we get $i + j \equiv 0 \bmod p$. This is impossible because i and j are positive integers less than or equal to $\frac{p-1}{2}$; hence their sum is positive and less than p. We conclude that $S \cap U = \varnothing$.

To show that $S \cup U = \left\{1, 2, \ldots, \frac{p-1}{2}\right\}$ notice first that by the definition of S we already have $S \subseteq \left\{1, 2, \ldots, \frac{p-1}{2}\right\}$. Also, any element of U has the form $p - r_k$ with $r_k \in T$. By definition of T, $r_k \in T$ means that $\frac{p-1}{2} < r_k < p$. Thus $1 \leq p - r_k < \frac{p-1}{2}$, so each element of U belongs to $\left\{1, 2, \ldots, \frac{p-1}{2}\right\}$; hence we also have $U \subseteq \left\{1, 2, \ldots, \frac{p-1}{2}\right\}$.

We now have $S \cup U \subseteq \left\{1, 2, \ldots, \frac{p-1}{2}\right\}$. We will now show that $S \cup U$ contains $\frac{p-1}{2}$ distinct elements, which will force our inclusion to be an equality. We proceed.

First, the integers $r_1, r_2, \ldots, r_{\frac{p-1}{2}}$ are all distinct. For if $r_i = r_j$, then $ai \equiv r_i = r_j \equiv aj \bmod p$; hence $(a, p) = 1$ implies $i \equiv j \bmod p$. Thus $i = j$ (By Lemma 4.9), showing that $r_1, r_2, \ldots, r_{\frac{p-1}{2}}$ are all distinct. Since S is a subset of $\left\{r_1, r_2, \ldots, r_{\frac{p-1}{2}}\right\}$, all the integers in S are distinct. Further, T is another subset, so all its elements are distinct, and this clearly means that the members of $U = \{p - r_k | r_k \in T\}$ are also distinct. In addition, $S \cap U = \varnothing$, so the integers in $S \cup U$ are all distinct. Finally, U and T have the same number of elements and since $S \cap T = \varnothing$, the number of elements in $S \cup U$ = number of elements in $S \cup T$. However, $S \cup T = \left\{r_1, r_2, \ldots, r_{\frac{p-1}{2}}\right\}$, which has exactly $\frac{p-1}{2}$ elements. ∎

We introduce two final pieces of useful notation. If A is a finite nonempty set of integers, we let $\prod A$ denote the product of all the integers in A, and we let $\sum A$ denote the sum of all the integers in A. In our example, if A is the set $T = \{10, 15, 13\}$, then

$$\prod A = \prod T = 10 \cdot 15 \cdot 13 = 1950$$

$$\sum A = \prod T = 10 + 15 + 13 = 38$$

We proceed to an easy lemma, keeping in mind that t denotes the number of elements in the set T.

Lemma 9.11

(i) $\prod U \equiv (-1)^t \prod T \bmod p$.
(ii) $\sum U = tp - \sum T$.

Proof.

(i) We have $U = \{p - r_k | r_k \in T\}$ and $p - r_k \equiv -r_k \bmod p$. Thus $\prod U = \prod \{p - r_k | r_k \in T\} \equiv \prod \{-r_k | r_k \in T\} \bmod p$. Now T contains t elements; hence $\prod \{-r_k | r_k \in T\}$ involves a factor of -1 occurring t times. Thus

$$\prod U \equiv \prod \{-r_k | r_k \in T\} \equiv (-1)^t \cdot \prod \{r_k | r_k \in T\} \equiv (-1)^t \prod T \bmod p$$

(ii) $\sum U = \sum \{p - r_k | r_k \in T\}$. Because T has t elements, there are t occurrences of p in $\{p - r_k | r_k \in T\}$. Thus $\sum U = tp - \sum \{r_k | r_k \in T\} = tp - \sum T$. ∎

The important step coming up is known as Gauss's Lemma; it relates the Legendre symbol $\left(\frac{a}{p}\right)$ to the sets S, T, and U.

Theorem 9.12. If p is an odd prime and $a \in \mathbb{Z}$ with $(a, p) = 1$, then $\left(\frac{a}{p}\right) = (-1)^t$.

Proof. For convenience, let $d = \frac{p-1}{2}$. As before, if $1 \le k \le d$, then $ka = p \cdot m_k + r_k$; hence $ka \equiv r_k \bmod p$. Then

$$d!\,a^d = (1 \cdot a)(2a) \cdots (da) \equiv r_1 r_2 \cdots r_d \bmod p$$

Since $S \cup T = \{r_1, r_2, \ldots, r_d\}$ and $S \cap T = \varnothing$, we have

$$(+) \qquad (\prod S)(\prod T) = \prod S \cup T = r_1 r_2 \cdots r_d$$

Thus

$$d!\,a^d \equiv (\prod S)(\prod T) \bmod p$$

From Lemma 9.11, we have $\prod U = (-1)^t \cdot \prod T$; multiplying through by $(-1)^t$ gives us

$$(-1)^t \cdot \prod U = (-1)^{2t} \cdot \prod T = \prod T$$

Substituting for $\prod T$ in $(+)$ results in

$$d!\,a^d \equiv (-1)^t (\prod S)(\prod U) \bmod p$$

Now by Lemma 9.10, $S \cup U = \{1, 2, \ldots, d\}$ and $S \cap U = \varnothing$. Therefore,

$$\prod S \cdot \prod U = 1 \cdot 2 \cdots d = d!$$

so

$$d!\,a^d \equiv (-1)^t d! \bmod p$$

Since $(d!, p) = 1$, we may cancel $d!$ in this congruence and arrive at

$$a^{\frac{p-1}{2}} = a^d \equiv (-1)^t \bmod p$$

By Euler's Criterion (Theorem 9.9), we already know that

$$\left(\frac{a}{p}\right) \equiv a^{\frac{p-1}{2}} \bmod p$$

so

$$\left(\frac{a}{p}\right) \equiv (-1)^t \bmod p$$

Since p is an odd prime and $\left(\frac{a}{p}\right) = \pm 1$, $(-1)^t = \pm 1$, we conclude that

$$\left(\frac{a}{p}\right) = (-1)^t. \qquad \blacksquare$$

Continuing in the development of the proof of the Quadratic Reciprocity Theorem, we next add to our earlier assumption that $(a, p) = 1$ the condition that a is an odd integer.

Lemma 9.13. Assume that $(a, p) = 1$ and a is odd. Then:

(i) For $k = 1, 2, \ldots, \frac{p-1}{2}$, we have $m_k \equiv k - r_k \bmod 2$.

(ii) If $m = m_1 + m_2 + \cdots + m_{\frac{p-1}{2}}$, then $m \equiv t \bmod 2$.

Proof.

(i) Recall that m_k is given by the relation

$$k \cdot a = p \cdot m_k + r_k, \qquad 1 \leq r_k < p$$

Adding $k(p - a)$ to both sides and simplifying, we get

$$(k - m_k)p = r_k + k(p - a)$$

Since p and a are both odd, $k(p - a)$ is even, so

$$(k - m_k)p \equiv r_k \bmod 2$$

Because $p \equiv 1 \bmod 2$ (p is odd), this reduces to

$$k - m_k \equiv r_k \bmod 2$$

so

$$m_k \equiv k - r_k \bmod 2$$

(ii) By part (i), we have

$$m \equiv (1 - r_1) + (2 - r_2) + \cdots + \left(\frac{p-1}{2} - r_{\frac{p-1}{2}}\right) \bmod 2$$

Thus

$$m \equiv \left(1 + 2 + \cdots + \frac{p-1}{2}\right) - \left(r_1 + r_2 + \cdots + r_{\frac{p-1}{2}}\right) \bmod 2$$

Now since $S \cup T = \left\{r_1, r_2, \ldots, r_{\frac{p-1}{2}}\right\}$ and $S \cap T = \varnothing$, clearly

$$r_1 + r_2 + \cdots + r_{\frac{p-1}{2}} = \sum S + \sum T$$

so

$$m \equiv \left(1 + 2 + \cdots + \frac{p-1}{2}\right) - \left(\sum S + \sum T\right) \bmod 2$$

However, we also have $S \cup U = \left\{1, 2, \ldots, \frac{p-1}{2}\right\}$ and $S \cap U = \varnothing$.

Therefore, $1 + 2 + \cdots + \frac{p-1}{2} = \sum S + \sum U$, so

$$m \equiv \left(\sum S + \sum U\right) - \left(\sum S + \sum T\right) \equiv \sum U - \sum T \bmod 2$$

By Lemma 9.11, we have $\sum U = pt - \sum T$, so

$$m \equiv \sum U - \sum T \equiv pt - \sum T - \sum T \equiv pt \bmod 2$$

Of course, $p \equiv 1 \bmod 2$, so that

$$m \equiv t \bmod 2 \qquad \blacksquare$$

Theorem 9.14. If p is an odd prime and $(a, p) = 1$, then

$$\left(\frac{a}{p}\right) = (-1)^m$$

where $m = m_1 + m_2 + \cdots + m_{\frac{p-1}{2}}$.

Proof. By (ii) of the previous lemma, $t \equiv m \bmod 2$ and hence t and m are both even or both odd. In either case, $(-1)^m = (-1)^t$. By Theorem 9.12,

$$\left(\frac{a}{p}\right) = (-1)^t = (-1)^m \qquad \blacksquare$$

This last theorem is certainly a bizarre result. Yet it is just what we need in order to reach our goal. The plan is to apply this theorem to two distinct odd primes p and q, first letting q play the role of a. If we write

$$1 \cdot q = m_1 p + r_1, 2q = m_2 p + r_2, \ldots, \left(\frac{p-1}{2}\right)q = m_{\frac{p-1}{2}} \cdot p + r_{\frac{p-1}{2}}$$

and let $m = m_1 + m_2 + \cdots + m_{\frac{p-1}{2}}$, then Theorem 9.14 tells us that

$$\left(\frac{q}{p}\right) = (-1)^m$$

Now we reverse the roles of p and q, writing

$$1 \cdot p = n_1 q + s_1, 2 \cdot p = n_2 q + s_2, \ldots, \left(\frac{q-1}{2}\right)p = n_{\frac{q-1}{2}} \cdot q + s_{\frac{q-1}{2}}$$

if $n = n_1 + n_2 + \cdots + n_{\frac{q-1}{2}}$, then Theorem 9.14 tells us that

$$\left(\frac{p}{q}\right) = (-1)^n$$

In order to find a relationship between $\left(\frac{q}{p}\right)$ and $\left(\frac{p}{q}\right)$, it now becomes clear that we should find a relationship between m and n. Gauss found such a relationship! Before reading Theorem 9.17, you may want to try some examples and see if you can discover what Gauss saw.

Lemma 9.15. Let $k \in \left\{1, 2, \ldots, \frac{p-1}{2}\right\}$ and let h be any integer. Then $hp - kq \neq 0$. Furthermore, if $h \leq m_k$, then $hp - kq < 0$, while if $h > m_k$, then $hp - kq > 0$.

Proof. We have $kq = m_k \cdot p + r_k$, where $1 \leq r_k < p$. Hence $hp - kq = (h - m_k)p - r_k$. Since $r_k \geq 1$, we know that $p \nmid r_k$, so $hp - kq \neq 0$. Now note that if $h \leq m_k$, then $h - m_k \leq 0$, so $(h - m_k)p - r_k < 0$. On the other hand, if $h > m_k$, then $h - m_k > 0$. Because $p > r_k$ we get $(h - m_k)p > (h - m_k)r_k \geq r_k$; hence $(h - m_k)p - r_k > 0$. ∎

Lemma 9.16. Let p and q be distinct odd primes. For $k = 1, 2, \ldots, \frac{p-1}{2}$, let $kq = m_k \cdot p + r_k$ with $1 \leq r_k < p$. Then $m_k \leq \frac{q-1}{2}$.

Proof. Since $k \leq \frac{p-1}{2}$, we have $kq \leq \frac{p-1}{2} \cdot q = \frac{pq}{2} - \frac{q}{2}$. Then

$$m_k \cdot p < m_k \cdot p + r_k = kq \leq \frac{pq}{2} - \frac{q}{2}$$

Thus

$$m_k \leq \frac{q}{2} - \frac{q}{2p} < \frac{q}{2}$$

That is, m_k is an integer less than $\frac{q}{2}$. However, q is an odd number, so the largest integer less than $\frac{q}{2}$ is $\frac{q-1}{2}$. It follows that $m_k \leq \frac{q-1}{2}$. ∎

Theorem 9.17. Let p and q be distinct odd primes and let

$$kq = m_k \cdot p + r_k \qquad \text{with} \quad 1 \leq r_k < p \quad \left(k = 1, 2, \ldots, \frac{p-1}{2}\right)$$

$$hp = n_h \cdot q + s_h \qquad \text{with} \quad 1 \leq s_h < q_1 \quad \left(h = 1, 2, \ldots, \frac{q-1}{2}\right)$$

If $m = m_1 + m_2 + \cdots + m_{\frac{p-1}{2}}$ and $n = n_1 + n_2 + \cdots + n_{\frac{q-1}{2}}$, then $m + n = \frac{p-1}{2} \cdot \frac{q-1}{2}$.

Proof. We will consider the following array of numbers:

$$\begin{array}{ccccc}
p - q & 2p - q & 3p - q & \cdots & \left(\frac{q-1}{2}\right)p - q \\
p - 2q & 2p - 2q & 3p - 2q & \cdots & \left(\frac{q-1}{2}\right)p - 2q \\
p - 3q & 2p - 3q & 3p - 3q & \cdots & \left(\frac{q-1}{2}\right)p - 3q \\
\vdots & \vdots & \vdots & & \vdots \\
p - \left(\frac{p-1}{2}\right)q & 2p - \left(\frac{p-1}{2}\right)q & 3p - \left(\frac{p-1}{2}\right)q & \cdots & \left(\frac{q-1}{2}\right)p - \left(\frac{p-1}{2}\right)q
\end{array}$$

Notice that there are $\frac{p-1}{2} \cdot \frac{q-1}{2}$ numbers, and they are all nonzero by Lemma 9.15.

We wish to count the number of negative entries in this array, so let us examine an arbitrary row. Such a row is determined by an integer k (the coefficient of q), where $1 \le k \le \frac{p-1}{2}$. By Lemma 9.16, $m_k \le \frac{q-1}{2}$, so the kth row looks like:

$$p - kq \quad 2p - kq \quad 3p - kq \quad \cdots \quad m_k p - kq \quad \cdots \quad \left(\frac{q-1}{2}\right)p - kq$$

By Lemma 9.15, the negative entries in this row are precisely the numbers $p - kq, 2p - kq, \ldots, m_k p - kq$, and there are m_k of these numbers. Thus for each row k, the number of negative entries is m_k; hence summing up the number of negative entries in each row shows that the number of negative entires in the array is $m_1 + m_2 + \cdots + m_{\frac{q-1}{2}} = m$.

Next we reverse the roles of p and q and consider the array:

$$\begin{array}{ccccc}
q - p & 2q - p & 3q - p & \cdots & \left(\frac{p-1}{2}\right)q - p \\
q - 2p & 2q - 2p & 3q - 2p & \cdots & \left(\frac{p-1}{2}\right)q - 2p \\
q - 3p & 2q - 3p & 3q - 3p & \cdots & \left(\frac{p-1}{2}\right)q - 3p \\
\vdots & \vdots & \vdots & & \vdots \\
q - \left(\frac{q-1}{2}\right)p & 2q - \left(\frac{q-1}{2}\right)p & 3q - \left(\frac{q-1}{2}\right)p & \cdots & \left(\frac{p-1}{2}\right)q - \left(\frac{q-1}{2}\right)p
\end{array}$$

The argument given for the first array applies here and shows that the number of negative entries in this array is $n_1 + n_2 + \cdots + n_{\frac{q-1}{2}} = n$

However, the two arrays are related in the following manner. The kth *column* of the second array is obtained by multiplying each entry in the kth *row* of the first array by -1. That is, the entry in row h, column k of the second array is the term $kq - hp$, while the entry in row k, column h of the first array is $hp - kq = (-1)(kq - hp)$:

$$\begin{array}{cccccccccc}
 & & \textit{Array I} & & & & & \textit{Array II} & & \\
 & & \text{column } h & & & & & \text{column } k & & \\
 & & \vdots & & & & & \vdots & & \\
\text{row } k & \cdots & hp - kq & \cdots & & \text{row } h & \cdots & kq - hp & \cdots \\
 & & \vdots & & & & & \vdots & &
\end{array}$$

It is now clear that the number of positive entries in array I equals the number of negative entries in array II and the number of positive entries in array II equals the number of negative entries in array I. Since array II has exactly n negative entries, array I must have exactly n positive entries. We already know that array I has m negative entries. As mentioned earlier, there are no zero entries in array I, so array I has exactly $m + n$ entries. But as observed at the beginning, the array has $\frac{p-1}{2} \cdot \frac{q-1}{2}$ terms and hence $m + n = \frac{p-1}{2} \cdot \frac{q-1}{2}$. ∎

At last we prove our main result, which is worth restating.

Theorem 9.18. (Quadratic Reciprocity). Let p and q be distinct odd primes. If $p \equiv 3 \bmod 4$ and $q \equiv 3 \bmod 4$, then $\left(\frac{p}{q}\right) = -\left(\frac{q}{p}\right)$. Otherwise, $\left(\frac{p}{q}\right) = \left(\frac{q}{p}\right)$.

Proof. From Theorem 9.14, we have

$$\left(\frac{q}{p}\right) = (-1)^m, \qquad \left(\frac{p}{q}\right) = (-1)^n$$

where $m = m_1 + m_2 + \cdots + m_{\frac{p-1}{2}}$ and $n = n_1 + n_2 + \cdots + n_{\frac{q-1}{2}}$. Thus

$$\left(\frac{q}{p}\right)\left(\frac{p}{q}\right) = (-1)^{m+n} = (-1)^{\frac{p-1}{2} \cdot \frac{q-1}{2}}$$

the last equality coming from Theorem 9.17.

If $p \equiv 3 \bmod 4$ and $q \equiv 3 \bmod 4$, then $\frac{p-1}{2}$ and $\frac{q-1}{2}$ are both odd, hence $(-1)^{\frac{p-1}{2} \cdot \frac{q-1}{2}} = -1$. Since $\left(\frac{p}{q}\right)$ and $\left(\frac{q}{p}\right)$ each have value ± 1, in order for the product $\left(\frac{p}{q}\right)\left(\frac{q}{p}\right)$ to be -1 we must have one factor $+1$ and the other -1. Thus $\left(\frac{p}{q}\right) = -\left(\frac{q}{p}\right)$.

Now suppose that $p \not\equiv 3 \bmod 4$ (the case $q \not\equiv 3 \bmod 4$ is similar). Then $p \equiv 1 \bmod 4$, so $\frac{p-1}{2}$ is even. Therefore, $\frac{p-1}{2} \cdot \frac{q-1}{2}$ is even. Thus

$$\left(\frac{q}{p}\right)\left(\frac{p}{q}\right) = (-1)^{\frac{p-1}{2} \cdot \frac{q-1}{2}} = 1$$

Thus $\left(\frac{q}{p}\right)$ and $\left(\frac{p}{q}\right)$ must both have value $+1$, or else they both have value -1.

In either case, $\left(\frac{p}{q}\right) = \left(\frac{q}{p}\right)$. ∎

Finally, the Quadratic Reciprocity Theorem can be conveniently restated as

Corollary 9.19. Let p and q be distinct odd primes. Then

$$\left(\frac{p}{q}\right)\left(\frac{q}{p}\right) = (-1)^{\frac{p-1}{2}\cdot\frac{q-1}{2}}$$

EXERCISES 9.A

1. Find the distinct (nonzero) quadratic residues of the following primes: 11, 19, 23, 29.
2. Using the results of Exercise 9.A.1, evaluate the following:

   (a) $\left(\frac{4}{19}\right)$ (b) $\left(\frac{5}{29}\right)$ (c) $\left(\frac{17}{23}\right)$

   (d) $\left(\frac{5}{29}\right)$ (e) $\left(\frac{3}{11}\right)$ (*f*) $\left(\frac{3}{29}\right)$
3. Verify Theorem 9.3(iii) by calculating $\left(\frac{15}{29}\right)$ in two ways. Repeat for $\left(\frac{6}{11}\right)$ and $\left(\frac{21}{23}\right)$.
4. Complete the proof of Theorem 9.3(iii).
5. Evaluate: $\left(\frac{2}{3}\right)$, $\left(\frac{2}{5}\right)$, $\left(\frac{2}{7}\right)$, $\left(\frac{2}{11}\right)$, $\left(\frac{2}{13}\right)$, $\left(\frac{2}{17}\right)$, $\left(\frac{2}{19}\right)$, and $\left(\frac{2}{23}\right)$.
   Can you see a pattern? (If not, see Theorem 9.7.)
6. Verify that Corollary 9.5 holds for each pair of $p = 7$, $a = 2$; $p = 11$, $a = 4$; and $p = 13$, $a = 7$.
7. Verify that quadratic reciprocity (Theorem 9.6) holds for each pair of $p = 5, q = 7$; $p = 5, q = 17$; $p = 7, q = 19$; and $p = 13, q = 5$.
8. Following the examples in the text, calculate:

   (a) $\left(\frac{7}{31}\right)$ (b) $\left(\frac{10}{19}\right)$ (c) $\left(\frac{10}{37}\right)$

   (d) $\left(\frac{143}{8243}\right)$ (e) $\left(\frac{30}{101}\right)$ (f) $\left(\frac{170}{1667}\right)$

9. Use Euler's Criterion (Theorem 9.9) to show that

$$\left(\frac{5}{17}\right) = -1, \quad \left(\frac{2}{31}\right) = 1, \quad \left(\frac{-1}{19}\right) = -1$$

In the following four exercises we wish you to explore the notation and results needed in the proof of the Quadratic Reciprocity Theorem. In these exercises let $p = 23$ and $a = 7$. Note that $\frac{p-1}{2} = 11$.

10. For $k = 1, 2, 3, \ldots, 10, 11$, complete the following table, where $7 \cdot k = m_k \cdot 23 + r_k$ and $1 \leq r_k < 23$.

k	1	2	3	4	5	6	7	8	9	10	11
m_k	0	0			1						
r_k	7	14			12						

11. From the information in the previous exercise, list the elements of S, T, and U and calculate the number t.

12. Verify numerically Lemmas 9.10, 9.11, 9.13, and Theorems 9.12, 9.14, 9.17, 9.18 using the data from Exercise 9.A.10.

13. Repeat Exercises 9.A.10 through 12 with $p = 7$ and $a = 23$.

14. Prove Corollary 9.19.

EXERCISES 9.B

1. Derive Theorem 9.3(iii) and Theorem 9.4 from Euler's Criterion.

The next four exercises develop a proof of Theorem 9.7.

2. Let $p > 2$ be prime. If b is the smallest integer such that $\frac{p-1}{2} < 2b$, then show that $\frac{p-1}{2} < 2b \leq \frac{p-1}{2} + 2$ and from this show that $\frac{p-1}{4} - 1 < b - 1 \leq \frac{p-1}{4}$.

3. If $p > 2$ is prime and $a = 2$, show that if t is the number of elements of T (as defined prior to Lemma 9.11), then $t = \frac{p-1}{2} - (b-1)$, with b as in Exercise 9.B.2.

4. From Exercises 9.B.2 and 3, conclude that $\frac{p-1}{4} \leq t < \frac{p-1}{4} + 1$.

5. From Exercise 9.B.4 show that if $p \equiv \pm 1 \bmod 8$, then t is even, while if $p \equiv \pm 3 \bmod 8$, then t is odd. Then use Gauss's Lemma (Theorem 9.12) to prove Theorem 9.7.

Table of Squares $(1 \leq N \leq 200)$

N	N^2	N	N^2	N	N^2	N	N^2
1	1	51	2601	101	10201	151	22801
2	4	52	2704	102	10404	152	23104
3	9	53	2809	103	10609	153	23409
4	16	54	2916	104	10816	154	23716
5	25	55	3025	105	11025	155	24025
6	36	56	3136	106	11236	156	24336
7	49	57	3249	107	11449	157	24649
8	64	58	3364	108	11664	158	24964
9	81	59	3481	109	11881	159	25281
10	100	60	3600	110	12100	160	25600
11	121	61	3721	111	12321	161	25921
12	144	62	3844	112	12544	162	26244
13	169	63	3969	113	12769	163	26569
14	196	64	4096	114	12996	164	26896
15	225	65	4225	115	13225	165	27225
16	256	66	4356	116	13456	166	27556
17	289	67	4489	117	13689	167	27889
18	324	68	4624	118	13924	168	28224
19	361	69	4761	119	14161	169	28561
20	400	70	4900	120	14400	170	28900
21	441	71	5041	121	14641	171	29241
22	484	72	5184	122	14884	172	29584
23	529	73	5329	123	15129	173	29929
24	576	74	5476	124	15376	174	30276
25	625	75	5625	125	15625	175	30625
26	676	76	5776	126	15876	176	30976
27	729	77	5929	127	16129	177	31329
28	784	78	6084	128	16384	178	31684
29	841	79	6241	129	16641	179	32041
30	900	80	6400	130	16900	180	32400
31	961	81	6561	131	17161	181	32761
32	1024	82	6724	132	17424	182	33124
33	1089	83	6889	133	17689	183	33489
34	1156	84	7056	134	17956	184	33856
35	1225	85	7225	135	18225	185	34225
36	1296	86	7396	136	18496	186	34596
37	1369	87	7569	137	18769	187	34969
38	1444	88	7744	138	19044	188	35344
39	1521	89	7921	139	19321	189	35721
40	1600	90	8100	140	19600	190	36100
41	1681	91	8281	141	19881	191	36481
42	1764	92	8464	142	20164	192	36864
43	1849	93	8649	143	20449	193	37249
44	1936	94	8836	144	20736	194	37636
45	2025	95	9025	145	21025	195	38025
46	2116	96	9216	146	21316	196	38416
47	2209	97	9409	147	21609	197	38809
48	2304	98	9604	148	21904	198	39204
49	2401	99	9801	149	22201	199	39601
50	2500	100	10000	150	22500	200	40000

Table of Primes Less Than 10,000 ($p = a^2 + b^2 + c^2 + d^2$)

p	a	b	c	d
2	1	1	0	0
3	1	1	1	0
5	2	1	0	0
7	2	1	1	1
11	3	1	1	0
13	3	2	0	0
17	4	1	0	0
19	4	1	1	1
23	3	3	2	1
29	5	2	0	0
31	5	2	1	1
37	6	1	0	0
41	5	4	0	0
43	5	4	1	1
47	6	3	1	1
53	7	2	0	0
59	7	3	1	0
61	6	5	0	0
67	8	1	1	1
71	7	3	3	2
73	8	3	0	0
79	7	5	2	1
83	9	1	1	0
89	8	5	0	0
97	9	4	0	0
101	10	1	0	0
103	10	1	1	1
107	9	5	1	0
109	10	3	0	0
113	8	7	0	0
127	11	2	1	1
131	11	3	1	0
137	11	4	0	0
139	11	4	1	1
149	10	7	0	0
151	11	5	2	1
157	11	6	0	0
163	12	3	3	1
167	11	6	3	1
173	13	2	0	0
179	13	3	1	0
181	10	9	0	0
191	13	3	3	2
193	12	7	0	0
197	14	1	0	0
199	14	1	1	1
211	13	5	4	1
223	14	5	1	1
227	15	1	1	0
229	15	2	0	0
233	13	8	0	0
239	15	3	2	1
241	15	4	0	0
251	15	5	1	0
257	16	1	0	0
263	15	6	1	1
269	13	10	0	0
271	15	6	3	1
277	14	9	0	0
281	16	5	0	0
283	16	5	1	1
293	17	2	0	0
307	17	4	1	1
311	17	3	3	2
313	13	12	0	0
317	14	11	0	0
331	17	5	4	1
337	16	9	0	0
347	17	7	3	0
349	18	5	0	0
353	17	8	0	0
359	18	5	3	1
367	19	2	1	1
373	18	7	0	0
379	19	4	1	1
383	19	3	3	2
389	17	10	0	0
397	19	6	0	0
401	20	1	0	0
409	20	3	0	0
419	20	3	3	1
421	15	14	0	0
431	19	6	5	3
433	17	12	0	0
439	19	7	5	2
443	21	1	1	0
449	20	7	0	0
457	21	4	0	0
461	19	10	0	0
463	21	3	3	2
467	21	5	1	0
479	21	6	1	1
487	22	1	1	1
491	21	7	1	0
499	21	7	3	0
503	22	3	3	1
509	22	5	0	0
521	20	11	0	0

Table (*Continued*)

p	a	b	c	d	p	a	b	c	d
523	21	9	1	0	857	29	4	0	0
541	21	10	0	0	859	29	4	1	1
547	23	4	1	1	863	29	3	3	2
557	19	14	0	0	877	29	6	0	0
563	23	5	3	0	881	25	16	0	0
569	20	13	0	0	883	29	5	4	1
571	23	5	4	1	887	29	6	3	1
577	24	1	0	0	907	29	8	1	1
587	24	3	1	1	911	30	3	1	1
593	23	8	0	0	919	30	3	3	1
599	23	6	5	3	929	23	20	0	0
601	24	5	0	0	937	24	19	0	0
607	23	7	5	2	941	29	10	0	0
613	18	17	0	0	947	29	9	5	0
617	19	16	0	0	953	28	13	0	0
619	24	5	3	3	967	31	2	1	1
631	25	2	1	1	971	31	3	1	0
641	25	4	0	0	977	31	4	0	0
643	25	4	1	1	983	31	3	3	2
647	25	3	3	2	991	31	5	2	1
653	22	13	0	0	997	31	6	0	0
659	25	5	3	0	1009	28	15	0	0
661	25	6	0	0	1013	23	22	0	0
673	23	12	0	0	1019	31	7	3	0
677	26	1	0	0	1021	30	11	0	0
683	25	7	3	0	1031	31	6	5	3
691	25	8	1	1	1033	32	3	0	0
701	26	5	0	0	1039	31	7	5	2
709	22	15	0	0	1049	32	5	0	0
719	26	5	3	3	1051	32	5	1	1
727	26	7	1	1	1061	31	10	0	0
733	27	2	0	0	1063	31	10	1	1
739	27	3	1	0	1069	30	13	0	0
743	27	3	2	1	1087	31	11	2	1
751	27	3	3	2	1091	33	1	1	0
757	26	9	0	0	1093	33	2	0	0
761	20	19	0	0	1097	29	16	0	0
769	25	12	0	0	1103	33	3	2	1
773	22	17	0	0	1109	25	22	0	0
787	28	1	1	1	1117	26	21	0	0
797	26	11	0	0	1123	33	5	3	0
809	28	5	0	0	1129	27	20	0	0
811	28	5	1	1	1151	33	7	3	2
821	25	14	0	0	1153	33	8	0	0
823	27	9	3	2	1163	33	8	3	1
827	28	5	3	3	1171	33	9	1	0
829	27	10	0	0	1181	34	5	0	0
839	27	10	3	1	1187	33	9	4	1
853	23	18	0	0	1193	32	13	0	0

Table (*Continued*)

p	a	b	c	d	p	a	b	c	d
1201	25	24	0	0	1559	39	6	1	1
1213	27	22	0	0	1567	39	6	3	1
1217	31	16	0	0	1571	39	7	1	0
1223	34	7	3	3	1579	39	7	3	0
1229	35	2	0	0	1583	39	7	3	2
1231	35	2	1	1	1597	34	21	0	0
1237	34	9	0	0	1601	40	1	0	0
1249	32	15	0	0	1607	39	9	2	1
1259	35	5	3	0	1609	40	3	0	0
1277	34	11	0	0	1613	38	13	0	0
1279	35	7	2	1	1619	40	3	3	1
1283	35	7	3	0	1621	39	10	0	0
1289	35	8	0	0	1627	40	5	1	1
1291	35	8	1	1	1637	31	26	0	0
1297	36	1	0	0	1657	36	19	0	0
1301	26	25	0	0	1663	39	9	6	5
1303	35	7	5	2	1667	40	7	3	3
1307	36	3	1	1	1669	38	15	0	0
1319	35	9	3	2	1693	37	18	0	0
1321	36	5	0	0	1697	41	4	0	0
1327	35	10	1	1	1699	41	4	1	1
1361	31	20	0	0	1709	35	22	0	0
1367	35	9	6	5	1721	40	11	0	0
1373	37	2	0	0	1723	41	5	4	1
1381	34	15	0	0	1733	38	17	0	0
1399	37	5	2	1	1741	30	29	0	0
1409	28	25	0	0	1747	41	8	1	1
1423	37	7	2	1	1753	32	27	0	0
1427	37	7	3	0	1759	41	7	5	2
1429	30	23	0	0	1777	39	16	0	0
1433	37	8	0	0	1783	42	3	3	1
1439	37	6	5	3	1787	41	9	5	0
1447	38	1	1	1	1789	42	5	0	0
1451	37	9	1	0	1801	35	24	0	0
1453	38	3	0	0	1811	41	11	3	0
1459	37	9	3	0	1823	42	7	3	1
1471	38	5	1	1	1831	42	7	3	3
1481	35	16	0	0	1847	42	9	1	1
1483	37	8	7	1	1861	31	30	0	0
1487	38	5	3	3	1867	43	4	1	1
1489	33	20	0	0	1871	43	3	3	2
1493	38	7	0	0	1873	33	28	0	0
1499	37	11	3	0	1877	41	14	0	0
1511	38	7	3	3	1879	43	5	2	1
1523	39	1	1	0	1889	40	17	0	0
1531	39	3	1	0	1901	35	26	0	0
1543	39	3	3	2	1907	43	7	3	0
1549	35	18	0	0	1913	43	8	0	0
1553	32	23	0	0	1931	43	9	1	0

Table (*Continued*)

p	a	b	c	d	p	a	b	c	d
1933	42	13	0	0	2311	47	10	1	1
1949	43	10	0	0	2333	43	22	0	0
1951	43	10	1	1	2339	48	5	3	1
1973	38	23	0	0	2341	46	15	0	0
1979	44	5	3	3	2347	48	5	3	3
1987	44	7	1	1	2351	47	9	6	5
1993	43	12	0	0	2357	41	26	0	0
1997	34	29	0	0	2371	48	7	3	3
1999	43	11	5	2	2377	44	21	0	0
2003	44	7	3	3	2381	35	34	0	0
2011	44	7	5	1	2383	47	13	2	1
2017	44	9	0	0	2389	42	25	0	0
2027	45	1	1	0	2393	37	32	0	0
2029	45	2	0	0	2399	47	10	9	3
2039	45	3	2	1	2411	49	3	1	0
2053	42	17	0	0	2417	49	4	0	0
2063	45	6	1	1	2423	49	3	3	2
2069	38	25	0	0	2437	49	6	0	0
2081	41	20	0	0	2441	40	29	0	0
2083	45	7	3	0	2447	49	6	3	1
2087	45	7	3	2	2459	49	7	3	0
2089	45	8	0	0	2467	49	8	1	1
2099	45	8	3	1	2473	48	13	0	0
2111	45	9	2	1	2477	46	19	0	0
2113	33	32	0	0	2503	50	1	1	1
2129	40	23	0	0	2521	36	35	0	0
2131	45	9	5	0	2531	49	11	3	0
2137	36	29	0	0	2539	49	11	4	1
2141	46	5	0	0	2543	50	5	3	3
2143	46	5	1	1	2549	50	7	0	0
2153	37	28	0	0	2551	50	7	1	1
2161	44	15	0	0	2557	46	21	0	0
2179	45	12	3	1	2579	49	13	3	0
2203	45	13	3	0	2591	50	9	3	1
2207	46	9	3	1	2593	48	17	0	0
2213	47	2	0	0	2609	47	20	0	0
2221	45	14	0	0	2617	51	4	0	0
2237	46	11	0	0	2621	50	11	0	0
2239	47	5	2	1	2633	43	28	0	0
2243	47	5	3	0	2647	51	6	3	1
2251	47	5	4	1	2657	49	16	0	0
2267	47	7	3	0	2659	51	7	3	0
2269	37	30	0	0	2663	51	7	3	2
2273	47	8	0	0	2671	51	6	5	3
2281	45	16	0	0	2677	39	34	0	0
2287	47	7	5	2	2683	51	9	1	0
2293	42	23	0	0	2687	51	9	2	1
2297	44	19	0	0	2689	40	33	0	0
2309	47	10	0	0	2693	47	22	0	0

Table (*Continued*)

p	a	b	c	d	p	a	b	c	d
2699	51	9	4	1	3089	55	8	0	0
2707	52	1	1	1	3109	47	30	0	0
2711	51	10	3	1	3119	55	9	3	2
2713	52	3	0	0	3121	40	39	0	0
2719	51	10	3	3	3137	56	1	0	0
2729	52	5	0	0	3163	56	5	1	1
2731	52	5	1	1	3167	55	9	6	5
2741	46	25	0	0	3169	55	12	0	0
2749	43	30	0	0	3181	45	34	0	0
2753	52	7	0	0	3187	56	7	1	1
2767	51	11	6	3	3191	55	11	6	3
2777	44	29	0	0	3203	56	7	3	3
2789	50	17	0	0	3209	53	20	0	0
2791	51	10	9	3	3217	56	9	0	0
2797	51	14	0	0	3221	55	14	0	0
2801	49	20	0	0	3229	50	27	0	0
2803	52	9	3	3	3251	57	1	1	0
2819	53	3	1	0	3253	57	2	0	0
2833	48	23	0	0	3257	56	11	0	0
2837	41	34	0	0	3259	57	3	1	0
2843	53	5	3	0	3271	57	3	3	2
2851	53	5	4	1	3299	57	7	1	0
2857	51	16	0	0	3301	49	30	0	0
2861	50	19	0	0	3307	57	7	3	0
2879	53	6	5	3	3313	57	8	0	0
2887	53	7	5	2	3319	57	6	5	3
2897	44	31	0	0	3323	57	8	3	1
2903	53	9	3	2	3329	52	25	0	0
2909	53	10	0	0	3331	59	9	1	0
2917	54	1	0	0	3343	57	9	3	2
2927	54	3	1	1	3347	57	9	4	1
2939	53	11	3	0	3359	57	10	3	1
2953	53	12	0	0	3361	56	15	0	0
2957	46	29	0	0	3371	57	11	1	0
2963	53	12	3	1	3373	58	3	0	0
2969	40	37	0	0	3389	58	5	0	0
2971	53	12	3	3	3391	58	5	1	1
2999	54	9	1	1	3407	58	5	3	3
3001	51	20	0	0	3413	58	7	0	0
3011	53	12	7	3	3433	52	27	0	0
3019	53	13	5	4	3449	43	40	0	0
3023	54	9	5	1	3457	44	39	0	0
3037	54	11	0	0	3461	50	31	0	0
3041	55	4	0	0	3463	58	9	3	3
3049	45	32	0	0	3467	57	13	7	0
3061	55	6	0	0	3469	45	38	0	0
3067	55	5	4	1	3491	59	3	1	0
3079	55	7	2	1	3499	59	4	1	1
3083	55	7	3	0	3511	59	5	2	1

Table (*Continued*)

p	*a*	*b*	*c*	*d*
3517	59	6	0	0
3527	59	6	3	1
3529	48	35	0	0
3533	58	13	0	0
3539	59	7	3	0
3541	54	25	0	0
3547	59	8	1	1
3557	49	34	0	0
3559	59	7	5	2
3571	59	9	3	0
3581	59	10	0	0
3583	59	10	1	1
3593	53	28	0	0
3607	59	11	2	1
3613	43	42	0	0
3617	44	41	0	0
3623	59	9	6	5
3631	59	11	5	2
3637	46	39	0	0
3643	60	5	3	3
3659	60	7	3	1
3671	59	10	9	3
3673	48	37	0	0
3677	59	14	0	0
3691	60	9	3	1
3697	49	36	0	0
3701	55	26	0	0
3709	53	30	0	0
3719	59	15	3	2
3727	61	2	1	1
3733	57	22	0	0
3739	61	4	1	1
3761	56	25	0	0
3767	61	6	3	1
3769	60	13	0	0
3779	61	7	3	0
3793	52	33	0	0
3797	46	41	0	0
3803	61	9	1	0
3821	61	10	0	0
3823	61	10	1	1
3833	53	32	0	0
3847	61	1	1	1
3851	61	11	3	0
3853	62	3	0	0
3863	62	3	3	1
3877	54	31	0	0
3881	59	20	0	0
3889	60	17	0	0
3907	61	13	4	1
3911	62	7	3	3
3917	61	14	0	0
3919	62	7	5	1
3923	61	12	7	3
3929	52	35	0	0
3931	61	13	5	4
3943	62	9	3	3
3947	61	15	1	0
3967	62	11	1	1
3989	58	25	0	0
4001	49	40	0	0
4003	63	5	3	0
4007	63	6	1	1
4013	62	13	0	0
4019	63	7	1	0
4021	50	39	0	0
4027	63	7	3	0
4049	55	32	0	0
4051	63	9	1	0
4057	59	24	0	0
4073	52	37	0	0
4079	63	10	3	1
4091	63	11	1	0
4093	58	27	0	0
4099	64	1	1	1
4111	63	9	6	5
4127	63	11	6	1
4129	60	23	0	0
4133	62	17	0	0
4139	64	5	3	3
4153	48	43	0	0
4157	59	26	0	0
4159	63	10	9	3
4177	64	9	0	0
4201	51	40	0	0
4211	64	9	5	3
4217	64	11	0	0
4219	64	11	1	1
4229	65	2	0	0
4231	65	2	1	1
4241	65	4	0	0
4243	65	4	1	1
4243	53	38	0	0
4259	65	5	3	0
4261	65	6	0	0
4271	65	6	3	1
4273	57	32	0	0
4283	65	7	3	0

Table (*Continued*)

p	a	b	c	d	p	a	b	c	d
4289	65	8	0	0	4729	52	45	0	0
4297	61	24	0	0	4733	58	37	0	0
4327	65	10	1	1	4751	67	15	6	1
4337	49	44	0	0	4759	67	15	6	3
4339	65	8	7	1	4783	69	3	3	2
4349	50	43	0	0	4787	69	5	1	0
4357	66	1	0	0	4789	55	42	0	0
4363	65	11	4	1	4793	68	13	0	0
4373	62	23	0	0	4799	69	6	1	1
4391	66	5	3	1	4801	65	24	0	0
4397	61	26	0	0	4813	67	18	0	0
4409	53	40	0	0	4817	56	41	0	0
4421	65	14	0	0	4831	69	6	5	3
4423	66	7	3	3	4861	69	10	0	0
4441	60	29	0	0	4871	69	10	3	1
4447	66	9	3	1	4877	61	34	0	0
4451	65	15	1	0	4889	67	20	0	0
4457	64	19	0	0	4903	70	1	1	1
4463	66	9	5	1	4909	70	3	0	0
4481	65	16	0	0	4919	70	3	3	1
4483	65	16	1	1	4931	69	13	1	0
4493	67	2	0	0	4933	62	33	0	0
4507	67	4	1	1	4937	64	29	0	0
4513	48	47	0	0	4943	70	5	3	3
4517	49	46	0	0	4951	70	7	1	1
4519	67	5	2	1	4957	69	14	0	0
4523	67	5	3	0	4967	70	7	3	3
4547	67	7	3	0	4969	60	37	0	0
4549	65	18	0	0	4973	67	22	0	0
4561	60	31	0	0	4987	69	15	1	0
4567	67	7	5	2	4993	62	32	0	0
4583	67	9	3	2	4999	70	9	3	3
4591	67	10	1	1	5003	69	15	4	1
4597	54	41	0	0	5009	65	28	0	0
4603	67	8	7	1	5011	69	15	5	0
4621	61	30	0	0	5021	70	11	0	0
4637	59	34	0	0	5023	70	11	1	1
4639	67	11	5	2	5039	70	11	3	3
4643	68	3	3	1	5051	71	3	1	0
4649	68	5	0	0	5059	71	4	1	1
4651	68	5	1	1	5077	71	6	0	0
4657	56	39	0	0	5081	59	40	0	0
4663	67	13	2	1	5087	71	6	3	1
4673	68	7	0	0	5099	71	7	3	0
4679	67	10	9	3	5101	51	50	0	0
4691	68	7	3	3	5107	71	8	1	1
4703	67	14	3	3	5113	53	48	0	0
4721	64	25	0	0	5119	71	7	5	2
4723	68	9	3	3	5147	71	9	5	0

Table (*Continued*)

p	*a*	*b*	*c*	*b*	*p*	*a*	*b*	*c*	*d*
5153	68	23	0	0	5573	58	47	0	0
5167	71	11	2	1	5581	66	35	0	0
5171	71	11	3	0	5591	74	9	5	3
5179	71	11	4	1	5623	74	11	5	1
5189	70	17	0	0	5639	75	3	2	1
5197	66	29	0	0	5641	75	4	0	0
5209	72	5	0	0	5647	75	3	3	2
5227	72	5	3	3	5651	75	5	1	0
5231	71	10	9	3	5653	73	18	0	0
5233	72	7	0	0	5657	61	44	0	0
5237	71	14	0	0	5659	75	5	3	0
5261	70	19	0	0	5669	65	38	0	0
5273	67	28	0	0	5683	75	7	3	0
5279	71	15	3	2	5689	75	8	0	0
5281	60	41	0	0	5693	62	43	0	0
5297	71	16	0	0	5701	74	15	0	0
5303	71	15	6	1	5711	75	9	2	1
5309	53	50	0	0	5717	71	26	0	0
5323	72	11	3	3	5737	56	51	0	0
5333	73	2	0	0	5741	70	29	0	0
5347	73	4	1	1	5743	75	10	3	3
5351	73	3	3	2	5749	57	50	0	0
5381	65	34	0	0	5779	76	1	1	1
5387	73	7	3	0	5783	75	11	6	1
5393	73	8	0	0	5791	75	11	6	3
5399	73	6	5	3	5801	76	5	0	0
5407	73	7	5	2	5807	75	13	3	2
5413	63	38	0	0	5813	73	22	0	0
5417	59	44	0	0	5821	75	14	0	0
5419	73	9	3	0	5827	76	7	1	1
5431	73	10	1	1	5839	75	14	3	3
5437	69	26	0	0	5843	76	7	3	3
5441	71	20	0	0	5849	68	35	0	0
5443	73	8	7	1	5851	76	7	5	1
5449	60	43	0	0	5857	76	9	0	0
5471	73	9	6	5	5861	70	31	0	0
5477	74	1	0	0	5867	76	9	3	1
5479	74	1	1	1	5869	62	45	0	0
5483	73	12	3	1	5879	75	15	5	2
5501	74	5	0	0	5881	75	16	0	0
5503	74	5	1	1	5897	76	11	0	0
5507	73	13	3	0	5903	75	15	7	2
5519	74	5	3	3	5923	76	11	5	1
5521	65	36	0	0	5927	75	17	3	2
5527	74	7	1	1	5939	77	3	1	0
5531	73	12	7	3	5953	57	52	0	0
5557	74	9	0	0	5981	59	50	0	0
5563	73	15	3	0	5987	77	7	3	0
5569	63	40	0	0	6007	77	7	5	2

Table (*Continued*)

p	*a*	*b*	*c*	*b*	*p*	*a*	*b*	*c*	*d*
6011	77	9	1	0	6397	59	54	0	0
6029	77	10	0	0	6421	70	39	0	0
6037	66	41	0	0	6427	80	5	1	1
6043	77	8	7	1	6449	80	7	0	0
6047	77	10	3	3	6451	80	7	1	1
6053	62	47	0	0	6469	63	50	0	0
6067	77	11	4	1	6473	68	43	0	0
6073	77	12	0	0	6481	80	9	0	0
6079	77	11	5	2	6491	80	9	3	1
6089	67	40	0	0	6521	80	11	0	0
6091	77	12	3	3	6529	65	48	0	0
6101	74	25	0	0	6547	80	11	5	1
6113	73	28	0	0	6551	79	15	9	2
6121	64	45	0	0	6553	72	37	0	0
6131	77	12	7	3	6563	81	1	1	0
6133	78	7	0	0	6569	80	13	0	0
6143	78	7	3	1	6571	81	3	1	0
6151	78	7	3	3	6577	81	4	0	0
6163	77	15	3	0	6581	70	41	0	0
6173	58	53	0	0	6599	81	6	1	1
6197	71	34	0	0	6607	81	6	3	1
6199	78	9	5	3	6619	81	7	3	0
6203	77	16	3	3	6637	61	54	0	0
6211	77	16	5	1	6653	62	53	0	0
6217	76	21	0	0	6659	81	9	4	1
6221	61	50	0	0	6661	81	10	0	0
6229	73	30	0	0	6673	63	52	0	0
6247	79	2	1	1	6679	81	10	3	3
6257	79	4	0	0	6689	80	17	0	0
6263	79	3	3	2	6691	81	11	3	0
6269	70	37	0	0	6701	74	35	0	0
6271	79	5	2	1	6703	81	9	6	5
6277	79	6	0	0	6709	78	25	0	0
6287	79	6	3	1	6719	81	11	6	1
6299	79	7	3	0	6733	82	3	0	0
6301	75	26	0	0	6737	76	31	0	0
6311	79	6	5	3	6761	80	19	0	0
6317	74	29	0	0	6763	81	12	7	3
6323	79	9	1	0	6779	81	13	7	0
6329	77	20	0	0	6781	75	34	0	0
6337	71	36	0	0	6791	82	7	3	3
6343	79	10	1	1	6793	67	48	0	0
6353	73	32	0	0	6803	81	15	4	1
6359	79	10	3	3	6823	82	9	3	3
6361	69	40	0	0	6827	81	16	3	1
6367	79	11	2	1	6829	77	30	0	0
6373	78	17	0	0	6833	68	47	0	0
6379	79	11	4	1	6841	80	21	0	0
6389	58	55	0	0	6857	61	56	0	0

Table (*Continued*)

p	a	b	c	d
6863	82	11	3	3
6869	62	55	0	0
6871	82	11	5	1
6883	81	15	9	4
6899	83	3	1	0
6907	83	4	1	1
6911	83	3	3	2
6917	79	26	0	0
6947	83	7	3	0
6949	82	15	0	0
6959	83	6	5	3
6961	81	20	0	0
6967	83	7	5	2
6971	83	9	1	0
6977	71	44	0	0
6983	83	9	3	2
6991	83	10	1	1
6997	74	39	0	0
7001	76	35	0	0
7013	82	17	0	0
7019	83	11	3	0
7027	83	11	4	1
7039	83	11	5	2
7043	83	12	3	1
7057	84	1	0	0
7069	75	38	0	0
7079	83	10	9	3
7103	83	14	3	3
7109	70	47	0	0
7121	64	55	0	0
7127	83	15	3	2
7129	80	27	0	0
7151	83	15	6	1
7159	83	15	6	3
7177	84	11	0	0
7187	84	11	3	1
7193	67	52	0	0
7207	83	17	5	2
7211	84	11	5	3
7213	83	18	0	0
7219	84	9	9	1
7229	85	2	0	0
7237	81	26	0	0
7243	85	4	1	1
7247	85	3	3	2
7253	82	23	0	0
7283	85	7	3	0
7297	76	39	0	0
7307	95	9	1	0
7309	78	35	0	0
7321	61	60	0	0
7331	85	9	5	0
7333	63	58	0	0
7349	82	25	0	0
7351	85	11	2	1
7369	85	12	0	0
7393	72	47	0	0
7411	85	13	4	1
7417	84	19	0	0
7433	68	53	0	0
7451	85	15	1	0
7457	76	41	0	0
7459	85	15	3	0
7477	86	9	0	0
7481	85	16	0	0
7487	86	9	3	1
7489	80	33	0	0
7499	85	16	3	3
7507	85	16	5	1
7517	86	11	0	0
7523	85	17	3	0
7529	77	40	0	0
7537	79	36	0	0
7541	71	50	0	0
7547	85	15	9	4
7549	85	18	0	0
7559	86	9	9	1
7561	75	44	0	0
7573	87	2	0	0
7577	64	59	0	0
7583	87	3	2	1
7589	65	58	0	0
7591	87	3	3	2
7603	87	5	3	0
7607	87	6	1	1
7621	86	15	0	0
7639	87	6	5	3
7643	87	8	3	1
7649	68	55	0	0
7669	87	10	0	0
7673	83	28	0	0
7681	84	25	0	0
7687	87	10	3	3
7691	87	11	1	0
7699	87	11	3	0
7703	87	11	3	2
7717	81	34	0	0
7723	87	12	3	1

Table (*Continued*)

p	a	b	c	d	p	a	b	c	d
7727	87	11	6	1	8219	89	17	3	0
7741	75	46	0	0	8221	90	11	0	0
7753	88	3	0	0	8231	90	11	3	1
7757	86	19	0	0	8233	77	48	0	0
7759	87	10	9	3	8237	86	29	0	0
7789	83	30	0	0	8243	89	15	9	4
7793	88	7	0	0	8263	90	9	9	1
7817	64	61	0	0	8269	90	13	0	0
7823	87	15	5	2	8273	88	23	0	0
7829	73	50	0	0	8287	91	2	1	1
7841	79	40	0	0	8291	91	3	1	0
7853	67	58	0	0	8293	78	47	0	0
7867	88	11	1	1	8297	91	4	0	0
7873	68	57	0	0	8311	91	5	2	1
7877	74	49	0	0	8317	91	6	0	0
7879	87	15	9	2	8329	75	52	0	0
7883	88	11	3	3	8353	87	28	0	0
7901	85	26	0	0	8363	91	9	1	0
7907	88	9	9	1	8369	88	25	0	0
7919	87	18	5	1	8377	76	51	0	0
7927	89	2	1	1	8387	91	9	5	0
7933	78	43	0	0	8389	90	17	0	0
7937	89	4	0	0	8419	91	11	4	1
7949	82	35	0	0	8423	91	9	6	5
7951	89	5	2	1	8429	77	50	0	0
7963	89	5	4	1	8431	91	11	5	2
7993	72	53	0	0	8443	91	12	3	3
8009	85	28	0	0	8447	91	11	6	3
8011	89	9	3	0	8461	90	19	0	0
8017	84	31	0	0	8467	92	1	1	1
8039	89	10	3	3	8501	74	55	0	0
8053	87	22	0	0	8513	92	7	0	0
8059	89	11	4	1	8521	85	36	0	0
8069	65	62	0	0	8527	91	14	7	1
8081	80	41	0	0	8537	91	16	0	0
8087	89	11	6	3	8539	92	7	5	1
8089	67	60	0	0	8543	91	15	6	1
8093	82	37	0	0	8563	92	9	3	3
8101	90	1	0	0	8573	82	43	0	0
8111	90	3	1	1	8581	66	65	0	0
8117	89	14	0	0	8597	89	26	0	0
8123	89	12	7	3	8599	91	17	5	2
8147	89	15	1	0	8609	80	47	0	0
8161	81	40	0	0	8623	91	18	3	3
8167	90	7	3	3	8627	92	9	9	1
8171	89	15	5	0	8629	90	23	0	0
8179	89	16	1	1	8641	71	60	0	0
8191	90	9	3	1	8647	91	19	2	1
8209	72	55	0	0	8663	93	3	2	1

Table (*Continued*)

p	a	b	c	d	p	a	b	c	d
8669	85	38	0	0	9091	95	8	1	1
8677	81	46	0	0	9103	95	7	5	2
8681	91	20	0	0	9109	78	55	0	0
8689	92	15	0	0	9127	95	10	1	1
8693	73	58	0	0	9133	93	22	0	0
8699	93	7	1	0	9137	71	64	0	0
8707	93	7	3	0	9151	95	11	2	1
8713	93	8	0	0	9157	79	54	0	0
8719	93	6	5	3	9161	85	44	0	0
8731	93	9	1	0	9173	73	62	0	0
8737	84	41	0	0	9181	91	30	0	0
8741	79	50	0	0	9187	95	12	3	3
8747	93	9	4	1	9199	95	13	2	1
8753	92	17	0	0	9203	95	13	3	0
8761	75	56	0	0	9209	80	53	0	0
8779	93	11	3	0	9221	95	14	0	0
8783	93	11	3	2	9227	96	3	1	1
8803	93	12	3	1	9239	95	14	3	3
8807	93	11	6	1	9241	96	5	0	0
8819	93	13	1	0	9257	76	59	0	0
8821	89	30	0	0	9277	94	21	0	0
8831	93	13	3	2	9281	95	16	0	0
8837	94	1	0	0	9283	96	7	3	3
8839	94	1	1	1	9293	77	58	0	0
8849	68	65	0	0	9311	95	15	6	5
8861	94	5	0	0	9319	95	17	2	1
8863	94	5	1	1	9323	96	9	5	1
8867	93	13	7	0	9337	96	11	0	0
8887	94	7	1	1	9341	85	46	0	0
8893	78	53	0	0	9343	95	17	5	2
8923	93	16	3	3	9349	95	18	0	0
8929	73	60	0	0	9371	96	11	5	3
8933	82	47	0	0	9377	79	56	0	0
8941	90	29	0	0	9391	95	19	2	1
8951	94	9	5	3	9397	71	66	0	0
8963	93	17	5	0	9403	96	13	3	3
8969	88	35	0	0	9413	97	2	0	0
8971	93	15	9	4	9419	97	3	1	0
8999	94	9	9	1	9421	86	45	0	0
9001	80	51	0	0	9431	97	3	3	2
9007	94	13	1	1	9433	93	28	0	0
9011	93	19	1	0	9437	91	34	0	0
9013	87	38	0	0	9439	97	5	2	1
9029	95	2	0	0	9461	94	25	0	0
9041	95	4	0	0	9463	97	7	2	1
9043	95	4	1	1	9467	97	7	3	0
9049	93	20	0	0	9473	97	8	0	0
9059	95	5	3	0	9479	97	6	5	3
9067	95	5	4	1	9491	97	9	1	0

Table (*Continued*)

p	a	b	c	b	p	a	b	c	d
9497	76	61	0	0	9767	98	9	9	1
9511	97	10	1	1	9769	88	45	0	0
9521	89	40	0	0	9781	90	41	0	0
9533	82	53	0	0	9787	97	19	4	1
9539	97	11	3	0	9791	98	13	3	3
9547	97	11	4	1	9803	99	1	1	0
9551	97	9	6	5	9811	99	3	1	0
9587	97	13	3	0	9817	99	4	0	0
9601	95	24	0	0	9829	98	15	0	0
9613	98	3	0	0	9833	92	37	0	0
9619	97	13	5	4	9839	99	6	1	1
9623	98	3	3	1	9851	99	1	1	0
9629	98	5	0	0	9857	89	44	0	0
9631	98	5	1	1	9859	99	7	3	0
9643	97	15	3	0	9871	99	6	5	3
9649	80	57	0	0	9883	99	9	1	0
9661	70	69	0	0	9887	99	9	2	1
9677	94	29	0	0	9901	99	10	0	0
9679	98	7	5	1	9907	99	9	5	0
9689	92	35	0	0	9923	99	11	1	0
9697	81	56	0	0	9929	85	52	0	0
9719	98	9	5	3	9931	99	11	3	0
9721	75	64	0	0	9941	71	70	0	0
9733	97	18	0	0	9949	90	43	0	0
9739	97	17	5	4	9967	99	11	6	3
9743	98	11	3	3	9973	82	57	0	0
9749	82	55	0	0					

Index

Al-Fakhri, 122
Algorithm
  division, 14
  Euclid's, 26, 29
Al-Karkhi of Baghdad, 121
Archimedes, 110
Arithmetic function, 69
Arithmetics, 121, 142

Babylonian mathematics, 119 *ff*
Bachet de Meziriac, 121, 134

Canonical representation, 37
Chinese Remainder Theorem, 90, 92
Compatible system, 93
Congruences
  linear, 83 *ff*
  polynomial, 112 *ff*
  quadratic, 101 *ff*
  system of, 85 *ff*
Congruent modulo *m,* 47

Dedekind, R., 120
Diophantine equation, 21, 120
Diophantos, 21, 120, 128, 134
Dirichlet, L., 122
Dirichlet's Box Principle, 130
Disquisitiones Arithmeticae, 122
Divides, 9, 12 *ff*
Division Algorithm, 14

Eratosthenes, Sieve of, 39
Euclid, 34, 73, 155
Euclid's
  Algorithm, 26, 29
  Theorem, 35
Euler, L., 55, 110, 122, 128, 134
Euler's
  ϕ-function, 56, 66 *ff*
  Criterion, 163
  Theorem, 57

Fermat, Pierre de, 40, 55, 58, 121, 123, 128, 134
Fermat
  factorization, 40, 44
  prime, 73
Fermat's
  Last Theorem, 73, 122, 128, 131
  Little Theorem, 54, 58, 59
Feynman, R., 119
Frame, J. S., 164
Function
  μ (n), 75
  ϕ (n), 55, 66 *ff*
  σ (n), 70
  τ (n), 70
  arithmetic, 69
  multiplicative, 69
  strongly multiplicative, 80

Gauss, Carl Friedrich, 61, 73, 110, 122
Gauss's Lemma, 168
Goldbach's Conjecture, 41
Grand Vizier of Baghdad, 121
Greatest common divisor (GCD), 16
Greek mathematics, 119

Hilbert, D., 138
Holtzmann (Xylander), 121

Ideal in $\mathbb{Z}$, 18
Indices, 151
Induction, Principle of, 6
Infinite descent, 58, 131
Integers
  addition of, 1
  multiplication of, 2
  order properties of, 3
Inverse modulo m, 50

Kaplansky, I., 122
Kraitchik, M., 73
Kummer, E., 122

Lagrange, J., 61, 122, 134
Least common multiple, 31
Legendre, A. M., 155
Legendre symbol, 155
Lehmer, D. H., 73
Liber Abaci, 121

$\mu(n)$, 75
Mersenne, M., 41, 73
Mersenne prime, 41, 73
Mobius Inversion Formula, 76
Multiplicative function, 69

Newton, I., 110, 120

Order of a modulo m, 52
Ore, O., 123

Palatine Anthology, 120
Pascal, B., 59
Paulsell, B., 124
Perfect number, 71
Pigeon-Hole Principle, 130
Pisano, L., 121
Prime
  Fermat, 73
  largest known, 34
  Mersenne, 44, 73
  relatively, 25
Prime factorization, 33
  existence of, 33
  uniqueness of, 35
Prime number, 33
Primitive root, 143
Principle
  Dirichlet's Box, 130
  Induction, 6
  Pigeon-Hole, 130
  Well-Ordering, 3
Pythagoras, 119, 120
Pythagorean Theorem, 119
Pythagorean triple, 123
  leg of, 123
  primitive, 125
Pythagoreans, 120

Quadratic
  residue mod p, 107
  nonresidue mod p, 107
Quadratic Reciprocity Theorem, 109, 159, 174, 175

Relatively prime, 25
  pairwise, 90
Rothbart, A., 124

$\sigma(n)$, 70
Sieve of Eratosthenes, 39
Solutions modulo m, 84
Square-free, 132
Strongly multiplicative function, 80
Sums of squares
  four squares, 134 *ff*
  three squares, 138
  two squares, 128 *ff*

$\tau(n)$, 70
Telegraph, 110
Thue, A., 130

Triangular number, 81
Twin-primes, 41

Waring, E., 138, 139
Waring's Problem, 122, 138

Well-Ordering, Principle of, 2, 3
Wilson, J., 61, 139
Wilson's Theorem, 60

Xylander (Holtzmann), 121